Noise and Vibration Measurement:

Prediction and Mitigation

Proceedings of a Symposium sponsored by
the Environmental Engineering Divison of
the American Society of Civil Engineers
in conjunction with the
ASCE Convention in Denver, Colorado

May 1-2, 1985

Edited by William A. Redl

Published by the
American Society of Civil Engineers
345 East 47th Street
New York, New York 10017-2398

Library of Congress Catalog Card No.: 85-70497
ISBN 0-87262-445-5
Manufactured in the United States of America.

PREFACE

The fifteen papers contained in this Proceedings represent a comprehensive view of three major sources of noise and vibration in our environment:

- o construction equipment and operations
- o transportation system operations
- o stationary source operations

The papers document both theoretical applications with respect to measurement, monitoring, prediction and mitigation, and case studies where certain techniques were employed to reduce noise and vibration for sensitive receptors.

These papers have been accepted for publication by the Proceedings Editor, and are eligible for discussion in the Journal of the Environmental Engineering Division and for ASCE Awards. I would like to thank each of the authors for their contribution to these Proceedings, which can serve as an important reference document for practitioners in the field and for the general public as an introduction to the inventory and assessment of noise and vibration.

William A. Reall, P.E., AICP
Livingston, New Jersey
February 21, 1985

CONTENTS

PREDICTION OF HIGHWAY CONSTRUCTION NOISE LEVELS*

by

William Bowlby[1], A.M. ASCE and Louis F. Cohn[2], M. ASCE

Abstract

A model for the prediction of noise levels from highway construction operations has been developed under sponsorship of the Federal Highway Administration (FHWA). Noise sources are geometrically represented as points, lines and areas, and noise barrier attenuation may also be analyzed. Built into the model and accompanying computer program is a data base for over fifty types of construction equipment and operations. Use of the interactive computer program is illustrated through an example problem.

While use of the model is not required, FHWA regulations do mandate that highway construction noise be addressed during the development of Federal-aid highway projects. The model and program may therefore be used by highway noise control engineers for environmental impact assessment and for design of noise abatement measures.

Introduction

The National Environmental Policy Act of 1969 (NEPA) required that the potential impacts of projects involving Federal funds be identified and studied. One such impact can be an increase in community noise levels. The Federal Highway Administration (FHWA) interpreted the requirements to cover not only the noise impact of a planned highway project, but the noise impact of the construction process itself.

In its highway noise regulation ("Procedures for Analysis and Abatement of Highway Traffic Noise and Construction Noise", commonly referred to as FHPM 7-7-3), FHWA states that in addition to analyzing

---

[1]Assistant Professor of Civil Engineering, Vanderbilt University, Box 90-B, Nashville, TN 37235

[2]Professor and Chairman, Department of Civil Engineering, Speed Scientific School, University of Louisville, Louisville, KY 40292

*As published in ASCE Journal of Construction Engineering, June 1983.

the potential impact of highway traffic, the following steps should be taken by the performing agency:

1. Identify "receptors" in the community that are sensitive to construction noise;

2. Determine mitigation measures for those impacted by construction noise, weighing cost and feasibility against benefits;

3. Incorporate the needed abatement measures into the plans and specifications for the project [11].

NEPA and FHPM 7-7-3 were technology-forcing, in that little was known about construction noise at the time of their implementation. Specifically, there were no impact criteria, no widely accepted abatement strategies, and no detailed knowledge of the extent of the construction noise problem. FHWA did not include construction noise level criteria in its standards, nor did it require specific analysis tools to be used. The depth of the analysis was left to the judgement of the performing agency, depending on the particular highway project. In most cases, the result was a very cursory analysis and specification of simple abatement methods.

Nonetheless, FHWA recognized that there would be cases, especially in completing many urban Interstate highways, where impacts could be severe and simple strategies would not suffice. It also recognized its leadership role in providing guidance and analysis tools to state highway agencies. It therefore sponsored the development of a comprehensive analytical model for predicting highway construction noise levels. The model, presented in this paper, permits detailed analysis of construction noise impact and mitigation strategy effectiveness [3,9].

Construction Noise Modelling

A major problem in trying to model construction noise is that there are many potential types of equipment that operate in different modes at a construction site. Equipment can operate in stationary, mobile and quasi-mobile modes (examples being, respectively, a pump, a truck, and a loader moving back and forth in a small area). A second major problem is that construction noise levels may be constant, cyclic or intermittent (examples being, respectively, an air compressor, an excavator loading trucks, and a truck on a haul road).

The first problem is addressed by dividing construction noise sources into three geometric types: point sources, line sources and area sources. The point source represents stationary and quasi-mobile operations operating over a small area. The line source represents operations that move in a linear fashion, staying on a fairly consistent path. Line operations are further divided into haul and nonhaul sources. Haul sources are equipment involved in earthmoving: trucks or scrapers. Nonhaul sources include operations such as paving, a push dozer working with scrapers, or a motor grader on a haul road. Area sources represent operations involving movement over an area too large to allow the noise to be represented as radiating from a point. Examples of area sources include dozers involved in clearing and grubbing and compactors working in a fill area.

The second problem -- noise level variation -- is addressed by using the "equivalent sound level", Leq in decibels, as the descriptor that the model would predict. Leq is defined as the steady state sound level over some time period that represents the same amount of acoustic energy as the time-varying level being studied.

Mathematically:

$$Leq = 10 \log \; 1/(t_2-t_1) \int_{t_1}^{t_2} 10^{L(t)} dt \qquad (1)$$

where t_1 and t_2 are the bounds of the time period under study and $10^{L(t)/10}$ is the acoustic energy from a source at time t.

This descriptor is chosen for several reasons. It has been shown to correlate well with expressed annoyance by people from other environmental noise sources [10]. It works equally well for constant, cyclic or intermittent sounds because it represents integration and averaging of the instantaneous level. It does not depend on when in the time period the noise even occurred, again because of the integration over the entire time period.

Finally, the concept permits the contributions from many different sources to be separately calculated and then easily combined into the total level that a receptor would hear. This combination process is accomplished by logarithmic summation, and is mathematically represented as follows:

$$(Leq)_{total} = 10 \log \sum_{i=1}^{n} 10^{(Leq)_i/10} \qquad (2)$$

where: $(Leq)_{total}$ is the total equivalent sound level from n sources and

$(Leq)_i$ is the equivalent sound level from the ith source.

One further item must be resolved before the Leq values may be computed: the length of the time period for the analysis. In the development of the highway construction noise model, it was decided to use an 8-hour period. Simply stated, all of the sounds reaching a receptor from the different operations occurring in an 8-hour period would be energy-averaged into one equivalent level for the 8-hour period. It was felt that there was too much variation in levels in the course of a workday to use a shorter period, and that use of a longer period would open the possibility of exceeding the duration of the workday, thus diluting the calculation by including non-construction periods in the integration. Therefore, the basic values that the model calculates at each receptor are, first, the 8-hour equivalent sound level, Leq(8h), from each source, and second, the total 8-hour equivalent sound level, $Leq(8h)_{total}$, from all sources.

Modelling "Unshielded" Noise

A brief discussion of how the various noise source types are modelled is presented in this section. For a full presentation of the mathematics of the model, the reader is referred to the version of the paper published in the ASCE Journal of Construction Engineering, June 1983.

For any source type, the Leq calculation process consists of computing a reference emission level at a distance of 50 feet from the source and then adjusting the reference level as needed to represent the problem being studied. This adjustment process permits the rather general formulation given in Equation (1) to be broken down into a series of logical calculation steps.

The reference level is the energy-averaged emission level in decibels over some time period, such as the equipment operation cycle. This "energy-averaging" is very similar in concept to the way the overall Leq is defined in Equation (1). For haul line sources, the reference level also depends on the travel speed of the equipment and its operating mode (acceleration, deceleration or cruise). As will be illustrated in the section titled Computer Program, the model contains reference levels for a variety of equipment types.

The adjustments made to the reference level include:

- usage factor;
- distance adjustment;
- length adjustment (for line sources);
- segment adjustment (for line and area sources);
- barrier attenuation.

The usage factor is a function of the percent of hours in an 8-hour workday that the equipment is in operation.

The distance adjustment accounts for the difference in noise level between the reference distance (50 feet (15.2m)) and the actual distance to the receptor. It is a function of the source type (basically, point versus line) and the type of ground cover: sound is attenuated more rapidly over soft ground than over packed ("hard") earth.

The length adjustment is a function of the length of the path over which a line source operates: the longer the path, the more time it takes to return past a receptor and, therefore, the lower the Leq at the receptor.

The segment adjustment is a function of the horizontal angle at a receptor between the ends of a line or area source. As this angle increases, the receptor is exposed to a "larger" source. This adjustment also depends on the type of ground over which the sound travels.

Modelling Noise Barrier Attenuation

To this point, the discussion has been for unshielded noise sources. However, shielding of sources often occurs on a construction site both as a noise abatement measure (e.g., noise barrier, presence of material stockpile) and as an unintentional consequence of the operations (e.g., fill embankment, existing mound in terrain).

Noise barriers for line and area construction sources are analyzed in the same manner as highway noise barriers. Highway noise barrier analysis is typically accomplished using the Kurze-Anderson method for attenuation of the sound from an "incoherent" line source [2,7]. This method is an adaption of Maekawa's formulation for shielding of a point source [8]. In effect, a point source is moved along the line (roadway) and its attenuation is computed at each point along the line. These attenuations are then combined to obtain the total attenuation of the line source. Shielding for construction point sources is directly based on Maekawa's formulation.

The barrier attenuation thus calculated is then subtracted from the unshielded Leq(8h) to get the shielded Leq(8h) at the receptor for a particular source. The Leq(8h) values for all sources are then combined via Equation (2) to get the total level at the receptor.

Computer Program

The highway construction noise model has been developed into a package of two computer programs called HINPUT and HICNOM [3,9]. HINPUT is an interactive program that requests input data from the user, makes initial acoustical and geometrical calculations, and prepares a data file that is used by HICNOM for the detailed acoustical calculations. HICNOM then produces a report file that contains the results of these calculations. The interactive data input and the two files are illustrated in the example problem presented in the next section.

While the earlier discussion on the model referred to "distances" from a receiver and "angles" to line source or barrier endpoints, the user provides geometrical data to the HINPUT program in terms of three-dimensional coordinates (X,Y,Z) of key points in the system, such as receiver locations or line source endpoints. HINPUT then performs the necessary distance and angle calculations. Use of a coordinate system simplifies data input when there are many sources or receivers, as permitted in the program. Up to ten receivers, ten point sources, six line sources, five area sources and three barriers may be defined for a given computer run. Further, each line source, area source, or barrier may be divided into up to nineteen, nine or four straight-line segments, respectively. Thus, a line path with curves or changes in profile may be modelled as a series of line segments, or an irregularly shaped area may be more accurately approximated.

In addition, for haul line sources, the user can instruct HINPUT to generate a series of points to represent a turn-around loop at the end of the line by specifying the type of loop and its radius. Figure 4 illustrates the "loop types" that may be represented. HINPUT will also compute average travel speeds on each loop segment based on the approach speed, location of the "stopping point", and the size of each segment.

Finally, HINPUT has a built-in data base for over fifty types of models of construction equipment. Table 1 illustrates the sixteen basic "equipment type" groups and the models contained in each group. In some instances, such as for equipment type CONCRETE, the model numbers represent different types of operations rather than equipment models. This data base includes the results of measurements made during development and validation of the model [4,5], as well as other data [6]. In addition to these pre-programmed equipment models, the user may temporarily define new models during a given computer run or permanently add them to the program, as long as the needed acoustical and operational data are known or can be approximated.

Example of Program Use

The following problem illustrates use of HINPUT and HICNOM for a hypothetical situation. A haul road is to be located so that it passes near three houses. The highway agency wishes to determine if the noise levels will be excessive, and if so, to determine feasible abatement methods. Figure 5 shows a sketch of the site. For the analysis, the engineer estimates that 20 trucks and 15 scrapers will pass the houses per hour for an 8-hour day.

Figure 6 shows the data input interaction with HINPUT, with user responses to HINPUT requests underlined. Note that because the type of trucks is not known, a model number of 4 for "nominal" truck type is used, while for the scrapers, a model number of 3 was used. The "loop type" of 0 and "stopping point" of 0 indicate to the program that the equipment are neither stopping nor turning around in the vicinity of the houses.

Figure 7 shows the output file produced by HINPUT, which is read and used by HICNOM, and Figure 8 shows the results file produced by HICNOM. In Figure 7, the first line of the output is the problem title and the subsequent sets of lines contain data on point, line and area sources, and noise barriers, respectively. This data includes user-supplied values, such as the items under the headings X, Y, Z, EX.ATT. (DB/DD) and DESCRIPTION; program-supplied values, such as the source sound frequency ("FREQ."); and program-calculated values, such as LEQ(REF) (reference level including usage factor and distance adjustment) and VEH.DENS. ("vehicle density", in vehicles/mile).

In Figure 8, the Leq(8h) for the three receivers are 70.3, 75.9, and 70.3 dBA, respectively. The engineer judges that these levels are not excessive and concludes that abatement is not warranted. If abatement did seem needed, a probable strategy would be a temporary noise barrier. The engineer could return to the program to test different barrier heights and locations to determine the best design.

Practical Applications

The highway construction noise prediction model may be used, with varying degrees of complexity, during the planning, design and construction of highway facilities. During planning and preliminary design, when the environmental impact assessment is ordinarily prepared, the model may be used as a general screening tool to anticipate potentially impacted receivers.

During final design, when the highway alignment and grades are resolved, the locations of structures are known, and the need for certain types of operations such as pile driving are known, the model permits a quantitative assessment of the impact at specific receiver locations. If problems are identified, the model may be used to test the effectiveness of different abatement strategies such as temporary noise barriers, equipment enclosures, equipment locations, location of haul roads and material stockpiles, restriction of work hours, and use of alternative equipment or methods. Special specifications may then be prepared for the most effective strategies, and incorporated into the project's bid documents.

Finally, during construction, the model may be used to help the engineer respond to citizen complaints on construction noise. The relative contributions of the various sources may be determined, and if necessary, abatement strategies may be evaluated and designed.

Summary and Conclusions

A model for the prediction of highway construction noise levels has been developed under the sponsorship of the Federal Highway Administration. The model computes the 8-hour equivalent sound level, Leq(8h), at one or more noise receptors near a construction site. A data base of over 50 types or models of construction equipment has been compiled for use in the calculations. Sources are geometrically represented as operating at a point, along a line or over an area, and noise barrier attenuation may also be determined.

The model may be used during the planning, design and construction of highway projects to determine potentially impacted areas, and to design and evaluate noise abatement measures where warranted.

TABLE 1

NOISE SOURCE DATA

SOURCE TYPE (I)	MODEL NO.* (2)	DESCRIPTION (3)
BACKHOE	1	NOMINAL # **
	2	CATERPILLAR, KOEHRING
	3	P & H
LOADER	1	NOMINAL #
	2	3-YARD CAPACITY
	3	5-YARD CAPACITY
	4	7-YARD CAPACITY
	5	10-YARD CAPACITY
COMPRESSOR	1	NOMINAL #
	2	STANDARD
	3	QUIET, DOORS OPEN
	4	QUIET, DOORS CLOSED
PILE DRIVER	1	NOMINAL #
	2	CURRENT DATA
PUMP	1	63 DB @ 50 FEET
	2	76 DB @ 50 FEET
	3	NOMINAL #
CRANE	1	NOMINAL #
	2	LOW
	3	MEDIUM
	4	HIGH
BREAKER	1	ROCK DRILL (NOMINAL) #
	2	STD. JACKHAMMER (NOMINAL)#
	3	MUFFLED JACKHAMMER

* THE ENGINEER MAY DEFINE ADDITIONAL MODEL NUMBERS IF SOURCE DATA ARE AVAILABLE.

\# USE THIS MODEL IF A GENERALIZED VALUE IS NEEDED.

** NOMINAL MEANS THAT THE DATA REPRESENTS AN AVERAGING OF DATA FROM PREVIOUS LITERATURE [6]

TABLE 1
NOISE SOURCE DATA (CONT'D)

SOURCE TYPE (I)	MODEL NO.* (2)	DESCRIPTION (3)
CONCRETE	1	CONCRETE POUR #
	2	NOMINAL BATCH PLANT
	3	BATCH PLANT
	4	PUMP #
	5	CEMENT MIXER #
GENERATOR	1	LOW LEVEL
	2	NOMINAL #
MISCELLANEOUS	1	GRINDER #
	2	CONCRETE SAW #
	3	FAN #
	4	WELDER (NOMINAL) # **
BULLDOZER	1	NOMINAL #
	2	CATERPILLAR D6, D7, D8
	3	CATERPILLAR D9
	4	D9 WITHOUT MUFFLER
GRADER	1	NOMINAL #
COMPACTOR	1	LOW
	2	NOMINAL #
	3	HIGH
PAVING	1	NOMINAL #
	2	CONCRETE PAVER
	3	ASPHALT PAVER
TRUCKS	1	10-YARD DUMP, QUIET
	2	10-YARD DUMP, NOISY
	3	DUAL 20-YARD TRAILERS
	4	NOMINAL #
SCRAPER	1	CATERPILLAR 631, MUFFLED
	2	CAT. 631, NO MUFF. (NOMINAL) #
	3	CATERPILLAR 623
	4	CATERPILLAR 636

* THE ENGINEER MAY DEFINE ADDITIONAL MODEL NUMBERS IF SOURCE DATA ARE AVAILABLE.
# USE THIS MODEL IF A GENERALIZED VALUE IS NEEDED.
** NOMINAL MEANS THAT THE DATA REPRESENTS AN AVERAGING OF DATA FROM PREVIOUS LITERATURE [6]

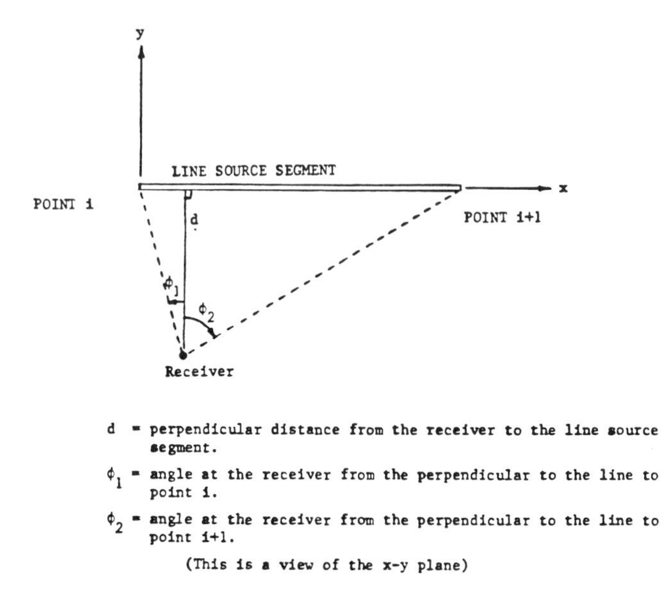

Figure 1. Geometry of Line Source Segment

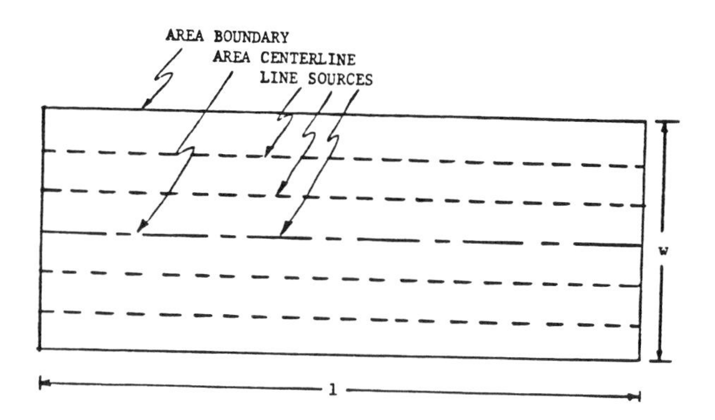

Figure 2. Area Source Representation

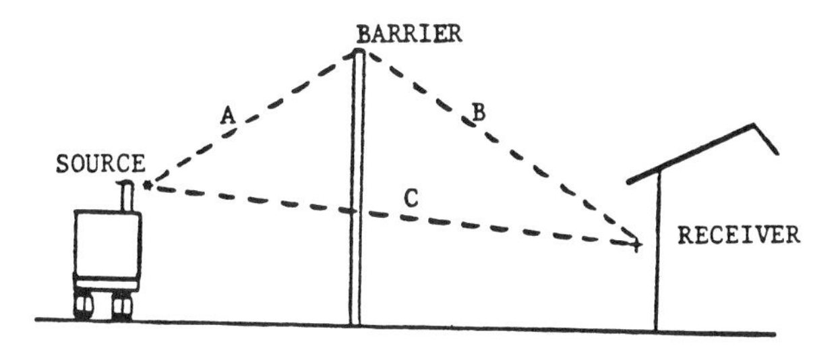

The quantity (A + B - C) is defined as the path length difference, δ, due to the presence of the barrier.

Figure 3. Path Length Difference Concept for Noise Barrier Attenuation

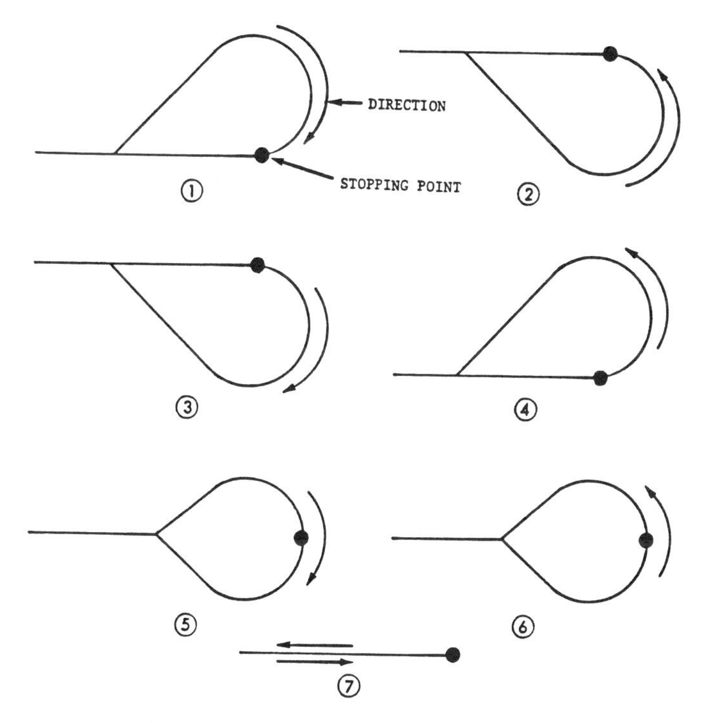

Figure 4. Haul Road Turn-around Loop Configurations

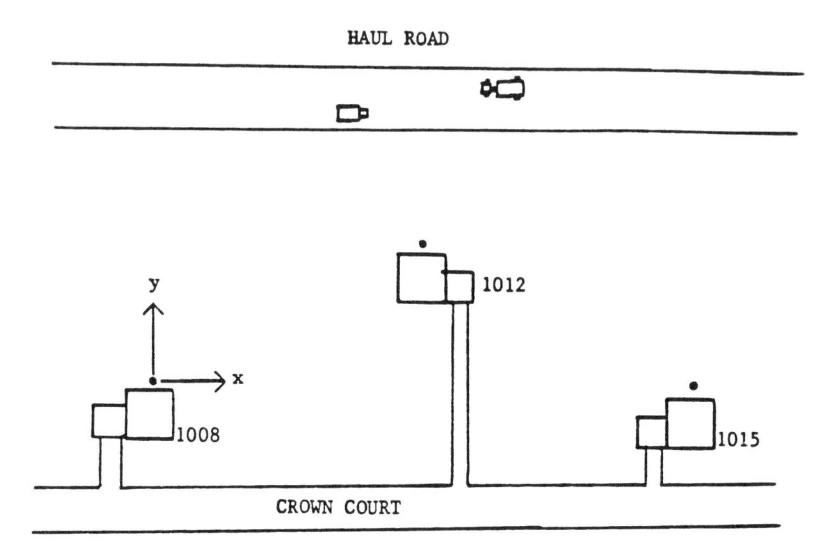

Figure 5. Sketch of Haul Road Example Problem

```
ENTER TITLE FOR THIS PROBLEM:
HAUL ROAD EXAMPLE: LEQ(8H) CALCULATIONS FOR 3 RECEIVERS
ENTER NUMBER OF RECEIVERS (MAXIMUM IS 10)
3
ENTER A DESCRIPTION OF RECEIVER # 1
(MAXIMUM OF 16 CHARACTERS - BLANK IF NONE)
1008 CROWN COURT
ENTER X, Y, Z AND EXCESS ATTENUATION (DB/DD) FOR  RECEIVER # 1
0 0 0 1.5
ENTER A DESCRIPTION OF RECEIVER # 2
(MAXIMUM OF 16 CHARACTERS - BLANK IF NONE)
1012 CROWN COURT
ENTER X, Y, Z AND EXCESS ATTENUATION (DB/DD) FOR  RECEIVER # 2
100 50 5 0.0
ENTER A DESCRIPTION OF RECEIVER # 3
(MAXIMUM OF 16 CHARACTERS - BLANK IF NONE)
1015 CROWN COURT
ENTER X, Y, Z AND EXCESS ATTENUATION (DB/DD) FOR  RECEIVER # 3
200 0 5 1.5
ENTER SOURCE TYPE - BLANK IF FINISHED

TRUCKS
ENTER A DESCRIPTION OF THE SOURCE
(MAXIMUM OF 16 CHARACTERS - BLANK IF NONE):
10-YD DUMPS
ENTER MODEL NUMBER (ENTER 0 TO DEFINE NEW MODEL NUMBER)
4

PROGRAM HAS AUTOMATICALLY ASSIGNED A GEOMETRY TYPE TO THIS SOURCE

ENTER HOURS WORKED DURING 8-HOUR DAY
(ENTER -1 TO COORDINATE THIS SOURCE"S PRODUCTION RATE WITH THAT OF
THE LAST PREVIOUSLY-ENTERED SOURCE HAVING A  PRODUCTION RATE)
8
LINE SOURCE:  ENTER NUMBER OF POINTS ON THE LINE, MAXIMUM IS 20
(ENTER 0 TO REUSE POINTS FROM LAST PREVIOUSLY-ENTERED LINE SOURCE)
2
ENTER X, Y, Z OF THE POINTS
-1000 100 0   1000 100 0
ENTER SPEED ON EACH SEGMENT (MPH)
45
ENTER VEHICLES PER HOUR (ONE WAY VOLUME)
20
ENTER TYPE AND RADIUS (FEET) OF RETURN LOOP
0 0
ENTER STOPPING POINT NUMBER
0
ENTER DECELERATION POINT NUMBER
0
```

Figure 6. Interactive Data Input for Example Problem
(User responses are underlined)

```
ENTER SOURCE TYPE - BLANK IF FINISHED

SCRAPER
ENTER A DESCRIPTION OF THE SOURCE
(MAXIMUM OF 16 CHARACTERS - BLANK IF NONE):
CAT 623
ENTER MODEL NUMBER (ENTER 0 TO DEFINE NEW MODEL NUMBER)
3

PROGRAM HAS AUTOMATICALLY ASSIGNED A GEOMETRY TYPE TO THIS SOURCE

ENTER HOURS WORKED DURING 8-HOUR DAY
(ENTER -1 TO COORDINATE THIS SOURCE"S PRODUCTION RATE WITH THAT OF
THE LAST PREVIOUSLY-ENTERED SOURCE HAVING A  PRODUCTION RATE)
8
LINE SOURCE:  ENTER NUMBER OF POINTS ON THE LINE, MAXIMUM IS 20
(ENTER 0 TO REUSE POINTS FROM LAST PREVIOUSLY-ENTERED LINE SOURCE)
0
ENTER SPEED ON EACH SEGMENT (MPH)
30
ENTER VEHICLES PER HOUR (ONE WAY VOLUME)
15
ENTER TYPE AND RADIUS (FEET) OF RETURN LOOP
0 0
ENTER STOPPING POINT NUMBER
0
ENTER DECELERATION POINT NUMBER
0
ENTER SOURCE TYPE - BLANK IF FINISHED


ENTER NUMBER OF BARRIERS (MAXIMUM IS 3)
0

ENTER THE NAME OF THE FILE FOR HICNOM USE;
MAXIMUM OF 10 (6.3) CHARACTERS, DEFAULT = HICNOM.DAT : EXAMP.DAT
```

Figure 6. Interactive Data Input for Example Problem (continued)

```
HAUL ROAD EXAMPLE: LEQ(8H) CALCULATIONS FOR 3 RECEIVERS

RECEIVER NUMBER          LEQ          DESCRIPTION


        1                 70.3       1008 CROWN COURT
        2                 75.9       1012 CROWN COURT
        3                 70.3       1015 CROWN COURT


COMPONENT CONTRIBUTIONS FOR RECEIVER NUMBER:  1

INDEX   INTENSITY     LEVEL      SOURCE           DESCRIPTION

  101   0.681668E+07  68.3     TRUCKS  1        10-YD DUMPS
  102   0.393301E+07  65.9     SCRAPER 1        CAT 623


COMPONENT CONTRIBUTIONS FOR RECEIVER NUMBER:  2

INDEX   INTENSITY     LEVEL      SOURCE           DESCRIPTION

  101   0.249010E+08  74.0     TRUCKS  1        10-YD DUMPS
  102   0.143671E+08  71.6     SCRAPER 1        CAT 623


COMPONENT CONTRIBUTIONS FOR RECEIVER NUMBER:  3

INDEX   INTENSITY     LEVEL      SOURCE           DESCRIPTION

  101   0.680725E+07  68.3     TRUCKS  1        10-YD DUMPS
  102   0.392757E+07  65.9     SCRAPER 1        CAT 623


KEY TO INDEX:
    X - POINT SOURCE, WHERE X OR XX IS INPUT SEQUENCE # OF POINT SOURCES.
   XX
  YXX - LINE SOURCE, WHERE XX IS INPUT SEQUENCE # OF LINE SOURCES
 YYXX                AND Y OR YY IS SEQUENCE # OF POINTS FOR THE XXTH LINE.
1YYXX - AREA SOURCE, WHERE XX AND YY ARE ANALAGOUS TO LINE SOURCE VARIABLES.
```

Figure 7. HINPUT Input Data File for Example Problem

```
Title           HAUL ROAD EXAMPLE: LEQ(8H) CALCULATIONS FOR 3 RECEIVERS
                  3  RECEIVERS
                     X          Y           Z      EX. ATT.(DB/DD)     DESCRIPTION
Receiver            0.0         0.0         0.0          1.50       1008 CROWN COURT
  Data            100.0        50.0         5.0          0.00       1012 CROWN COURT
                  200.0         0.0         5.0          1.50       1015 CROWN COURT
Point Data        0  POINT SOURCES
                  2  LINE SOURCES
                  2    POINTS             FREQ.:  500           TRUCKS  1        10-YD DUMPS
                     X          Y           Z      LEQ(REF) VEH. DENS.
                 -1000.0      100.0         6.0       92.9 0.0000840
                  1000.0      100.0         6.0        0.0 0.0000000
Line Source
  Data            2    POINTS             FREQ.:  500           SCRAPER 1        CAT 623
                     X          Y           Z      LEQ(REF) VEH. DENS.
                 -1000.0      100.0         6.0       90.0 0.0000945
                  1000.0      100.0         6.0        0.0 0.0000000
Area Data         0  AREA SOURCES
Barrier Data      0  BARRIERS
```

Figure 8. HICNOM Results File for Example Problem

APPENDIX I - REFERENCES

1. Barnes, J.D., et al., "Power Plan Construction Noise Guide," Bolt, Beranek and Newman, Inc. Report No. 3321, for the Empire State Electric Energy Research Corporation, 1977.

2. Barry, T.M., and Reagan, J.A., "FHWA Highway Traffic Noise Prediction Model," Report No. FHWA-RD-77-108, Federal Highway Administration, Washington, D.C., 1978.

3. Bowlby, W. and Cohn, L.F., "Highway Construction Noise - Environmental Assessment and Abatement, Volume IV - User's Manual for FHWA Highway Construction Noise Computer Program, HICNOM," Vanderbilt University Report VTR 81-2, for Federal Highway Administration, Washington, D.C., 1982.

4. Bowlby, W. and Cohn, L.F., "Highway Construction Noise - Environmental Assessment and Abatement, Volume V - Examples of Highway Construction Site Data: Measurements and Analysis," Vanderbilt University Report VTR 82-4, for Federal Highway Administration, Washington, D.C., 1982.

5. Fuller, W.R. and Brown, R., "Highway Construction Noise Field Measurements," Wyle Research Report No. WR 81-19, Appendices A-D, for Federal Highway Administration, Washington, D.C., 1981.

6. Fuller, W.R., et al., "Literature Review: Highway Construction Noise," Wyle Research Report WR 79-3, for Federal Highway Administration, Washington, D.C., 1979.

7. Kurze, U.J. and Anderson, G.S., "Sound Attenuation by Barriers," J. Applied Acoustics, Vol. 4, No. 1, 1971, pp. 35-53.

8. Maekawa, Z.E., "Noise Reduction by Screens," Applied Acoustics, Vol. 1, 1971, pp. 157-173.

9. Plotkin, K.J., "A Model for the Prediction of Highway Construction Noise," Wyle Research Report WR 80-58, for Federal Highway Administration, Washington, D.C., 1980.

10. ----, "Information on Levels of Environmental Noise Requisite to Protect Public Health and Welfare with an Adequate Margin of Safety," Report No. 550/9-74-004, U.S. Environmental Protection Agency, Washington, D.C., 1974.

11. ----, "Procedures for Abatement of Highway Traffic Noise and Construction Noise," Federal-aid Highway Program Manual, Volume 7, Chapter 7, Section 3, Federal Highway Administration, Washington, D.C., 1976.

NOISE AND AIR BLAST FROM CONSTRUCTION BLASTING

by

Charles H. Dowding
Associate Professor of Civil Engineering
Northwestern University
Evanston, IL 60201

ABSTRACT

Noise generated by construction blasting differs from other construction noise because of its frequency content. The audible high frequency portion is accompanied by a non audible, lower frequency portion that excites structures and in turn causes a secondary and audible rattle within a structure. Blasting noise occurs only a few times a day and lasts only one second or less. Since blasting is not continuous, human response cannot be extrapolated easily from studies of continuous sources such as traffic.

This paper presents the characteristics of blasting noise as well as its effects on structures and humans. The variation of measurement systems and the influence of weather on propagation relations will also be discussed.

AIR BLASTS AND REASONS FOR STUDY

Both the audible and inaudible air pressure waves generated by explosions are called air blasts. Just as with ground vibrations, these pressure waves can be described with time histories where the amplitude is air pressure instead of particle velocity. The higher frequency portion of the pressure wave is audible and is the sound that accompanies a blast; the lower frequency portion is not audible, but excites structures and in turn causes a secondary and audible rattle within a structure.

Air blasts are of interest for two reasons. First, they may crack windows; however, air blast pressures alone would have to be unusually high; 7 mbar, (0.1 psi, or 150 dB) to cause cracking of all windows. The second reason, human reaction, is the most intriguing, yet the most difficult to analyze quantitatively.

Previous researchers, Kamperman and Nicholson (1970), and Borsky (1965) have found that noise within a structure (from blasting and sonic booms, respectively) is the source of many complaints, even at very low levels of ground particle velocity. It appears that building motions, which are induced by air blasts, rattle loose objects within the structure which then startle the occupants.

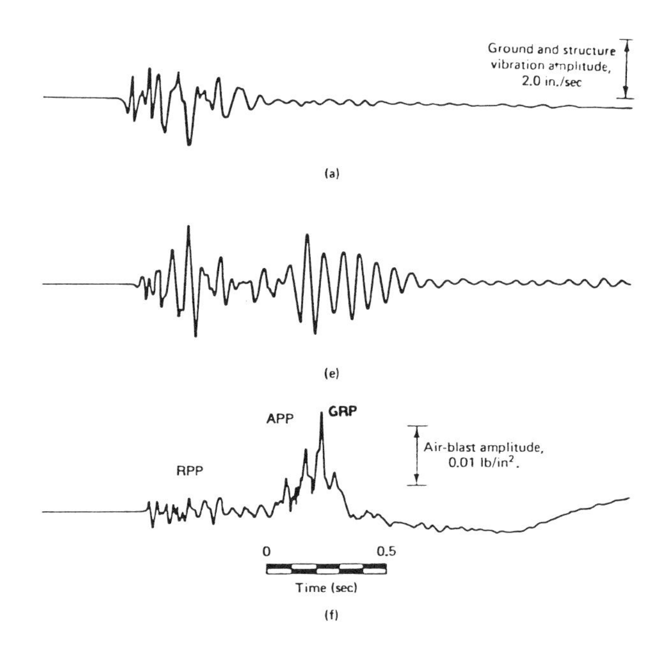

FIGURE 1

Air Blast Causes Relatively Large Wall Motions as Shown By Comparison of Outside Air Blast (f) with the Ground Vibrations (a) and the Wall Vibrations (e) (From Siskind et al. 1980a)

Time histories of structural response to ground vibration and air blast from a coal mining (high wall) shot are shown in Figure 1. This response was measured 218 m (716 ft) from a 1000 kg (2,200 lb) maximum charge per delay blast with a total charge weight of 11,770 kg (25,900 lb). The separate response of the structure to both the ground vibration and the air blast can be seen by comparing the air blast outside (f) and ground vibrations (a) with wall response (e). Obviously, the air blast could have contributed to the rattle heard inside the structure.

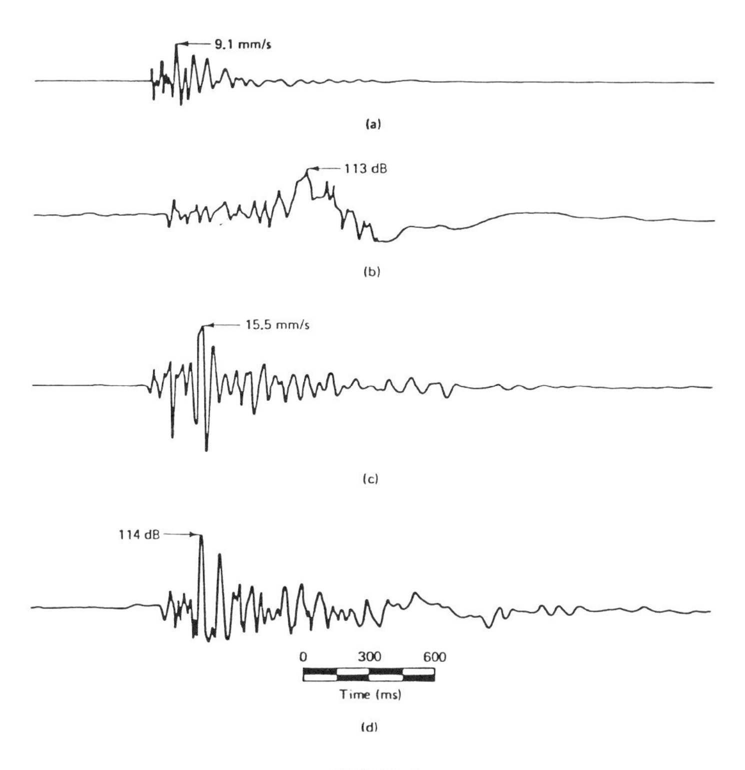

FIGURE 2

Inside Air Pressures Are More Intense and Can Be Generated by Ground Motions as Shown By Comparison of Inside (d) and Outside (b) Air Pressures and Floor (c) and Outside (a) Ground Motions (From Siskind et al. 1976)

This tendency for increased noise inside can be seen by comparing the inside and outside air blast pressures in Figure 2. The sound pressure peaks measured inside the structure (d) coincide with the floor motion (c) generated by the ground vibration (a) rather than the outside sound pressure (b) and are larger in amplitude. These motions were recorded 274 m (900 ft) from a 136 kg (300 lb) per delay quarry blast.

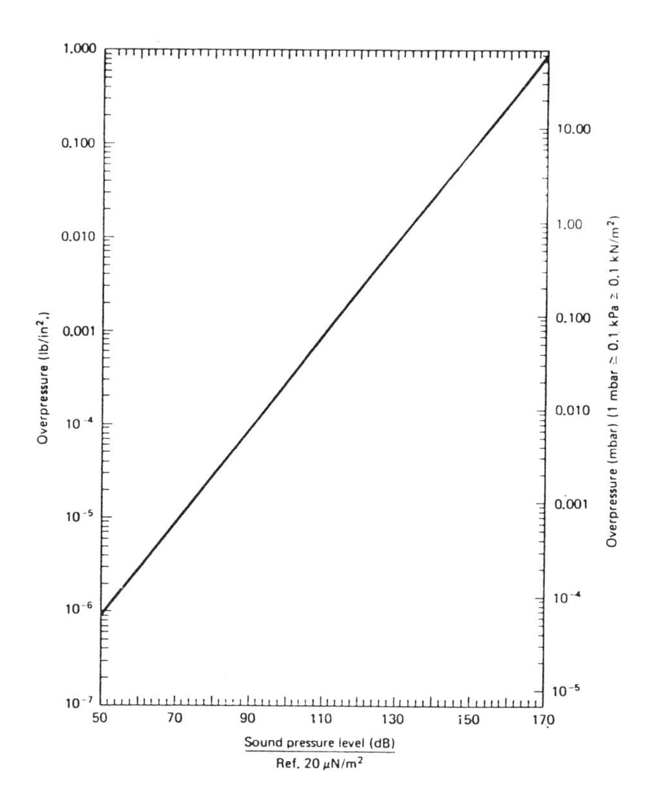

FIGURE 3

Translation of Decibels (dB) to Pressure (from Siskind et al. 1980a)

AIR BLAST MEASUREMENT AND CHARACTER

Sound level is reported with two distinct units of measure, pressure or decibels (dB). Pressure is the phenomenon that is measured by any sound or air blast transducer. Traditionally, sound has been reported by acoustical engineers in terms of decibels because of the wide range of amplitudes and frequencies that are detectable by the human ear. As shown in Figure 3, the decibel scale compresses the 10^{-7} to 10^{0} range of pressure amplitudes to a range of 50 to 170. Thus decible levels can be plotted on an arithmetic rather than a logarithmic scale. When pressure units are reported they are often called <u>overpressures</u> to indicate that the measured pressure is that above atmospheric.

Sound pressure is translated in the decibel scale by

$$dB = 20 \log_{10}\left(\frac{P}{P_o}\right) \qquad (1)$$

where P is the measured peak sound pressure and P_o is a reference pressure of $20 \times 10^{-6} N/m^2 (P_a)$ or $2.9 \times 10^{-9} lb/in.^2$. This conversion equation is plotted in Figure 3 so that sound pressure (greater than ambient or overpressure) can be directly translated to decibels, or vice versa.

The air blast (measured inside) from the quarry shot in Figure 2 is compared with that of a sonic boom and thunder in Figure 4. The quarry air blast and thunder are less intense than the sonic boom, but all three have similar dominant frequencies. The predominate peaks occur at frequencies of approximately 15, 17 & 13 H_z for quarrying, thunder and sonic boom (if only the rise time of the sonic boom is considered). Note the larger blast in Figure 1 also involves a much lower frequency (1 to 2 H_z) component. The higher-frequency spikes of the mining air blast in Figure 1 are related to the millisecond-delayed partitions of the blast.

The comparisons in Figure 4 show how misleading the decibel system is for those unaccustomed to its logarithmic nature. The pressures are plotted arithmetically in terms of Pascals, Pa, where 1 Pa is 1 KN/m^2 or 0.14 psi. Despite the factor of 6 difference between

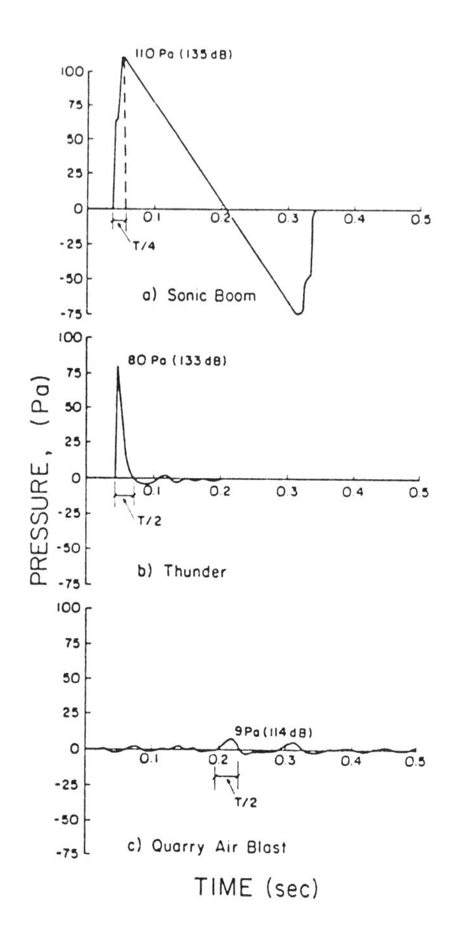

FIGURE 4

Comparison of Time Histories from A Sonic Boom, Thunder and A Quarry Blast Shows a Similarity in Principal Period (T) and Thus Frequency (f = 1/T)

the peak air pressures from sonic boom and quarry blast, there is only a 15% difference in terms of decibels.

Three main components of air blasts have been observed (Wiss and Linehan, 1978; Siskind et al., 1980a) and result from: (1) direct rock displacement at the blast, called the air pressure pulse (APP); (2) vibrating ground some distance from the blast, called the rock pressure pulse (RPP); and (3) venting at the hole caused by blowouts, called the gas release pulse (GRP). An APP is produced by each blast hole. Thus each delay can produce its own spike, as shown in Figure 1. At large distances the individual delay spikes begin to grow together as they are beginning to do in Figure 1. An RPP is generated by the vertical vibrations of the ground and arrives simultaneously with the ground vibration. As can be seen in the first portion of the air blast in Figure 1, the RPP is smaller than the APP in the second portion and will form the lower bound of the blast induced sound pressure.

The height of the individual spikes is controlled by the GRP and is dependent on the possibility of early venting of explosion gas pressures, called blowouts. These blowouts are usually responsible for unexpectedly high blast noise (air blasts). The venting often results from inadequate filling of the tops of blast holes with stemming material. One of the most common and most controllable sources of GRP's is uncovered detonation chord.

EFFECT OF WEIGHTING SCALES

Sound pressures are measured with a variety of instruments that do not respond equally to both the high-frequency pressures called sound and the low-frequency pulses (1 to 30 Hz) that excite structures. Figure 5 presents the frequency spectra or the response of transducers for the three commonly employed weighting scales. Air pressure fluctuations that occur at frequencies below 50 Hz are recorded below their field value by systems that employ "A" or "C" weighting scales. Although adequate for studying hearing response, A and C scale systems will not record the information necessary for correlation with structural response.

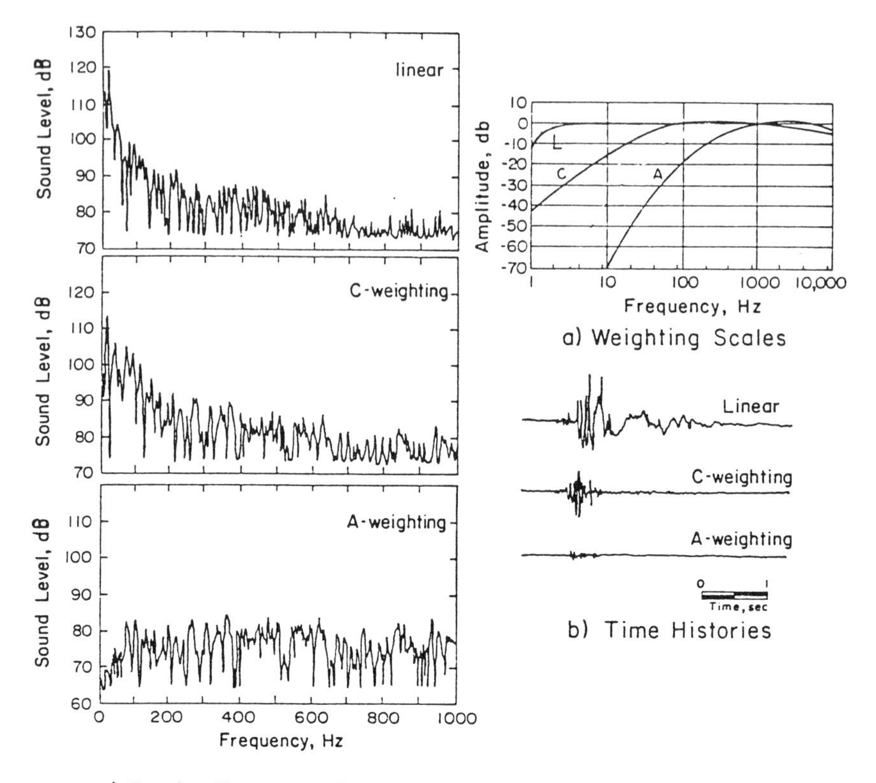

FIGURE 5

Since Weighted Filtering Affects the Measured Peaks and Time Histories, Only "Linear" Systems Should Be Employed

Such misrecording of low-frequency pressure pulses is shown in Figure 5c by the Fourier frequency spectra associated with pressure measured some 137 m (450 ft) from a quarry blast. The A and C weighting systems obtain amplitudes for the critical pressures (in the range 5 to 20 Hz) which are 55 dB & 10 dB too low at 20 Hz, respectively, and are even lower at 5 Hz. "Linear" systems that record accurately the pressures in both the structurally critical range (5 to 20 Hz) and the range critical for human hearing are necessary in critical blast monitoring situations. Perfectly undistorted records for research should be obtained with systems that are linear below 1 Hz.

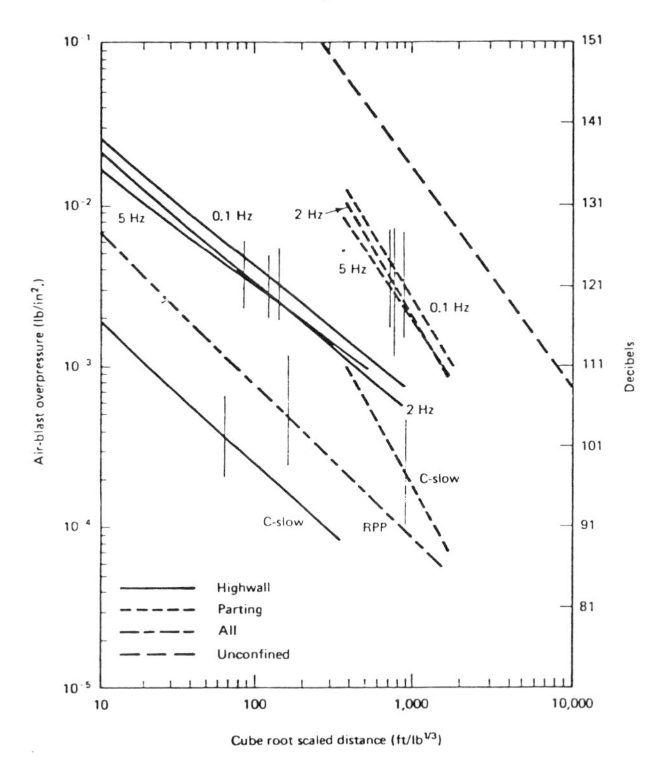

FIGURE 6

Affect of Weighting Scales on Air Blast Propagation Relations Shown By Comparison of C Slow and "Linear" (0.1 - $5H_z$) Measurements of High Wall Surface Mining Blasts (from Siskind et al. 1980a)

PROPAGATION AND SCALING OF AIR BLASTS

Propagation of air blasts has been studied by numerous investigators and is generally reported with cube root scaled distances. Cube root scaled distance is $R/W^{1/3}$ where R is the distance between shot and transducer and W is the maximum weight of explosives detonated at any instant or delay (Dowding, 1984). Probably the best summary of blast-induced air pressures is presented by Siskind et al. (1980a) and the following material is summarized directly from that work.

Figure 6 summarizes the effect of two important instrumentation and shot variables. First, the effect of the weighting scales is dramatically evident. C weighting dramatically reduces the recorded peak pressure from that recorded by a "linear" system at any scaled distance. This does not mean the peak is reduced by changing instruments, but that the C weighting system does not respond to the low--frequency pressure pulses. These pressure peaks excite structures and occupants whether or not they are measured. The 5-, 2-, 0.1-Hz labels denote the lower-frequency bounds of the recording capabilities of the "linear" systems. Second, the effect of venting caused by inadequate stemming can be observed by the higher than average pressures produced by the parting shots at any scaled distance. Parting shots are detonated in thin rock layers between coal strata in surface mines. Consequently, there is less hole height available for stemming and these shots many times blow out the stemming and result in abnormally high GRP air blast pressures. The unconfined relationship should be used for demolition of structures after modification for effects of weather and ground reflection.

Figure 7 shows the difference between air blasts produced by different industries at the same scaled distances and includes results of two studies. Both studies involved transducers that recorded linearly the low-frequency portion of the air blast. Although results of the two studies differed, they both indicated that in quarrying and construction the need to more finely fragment rock involves less confinement and results in larger air blast pressures.

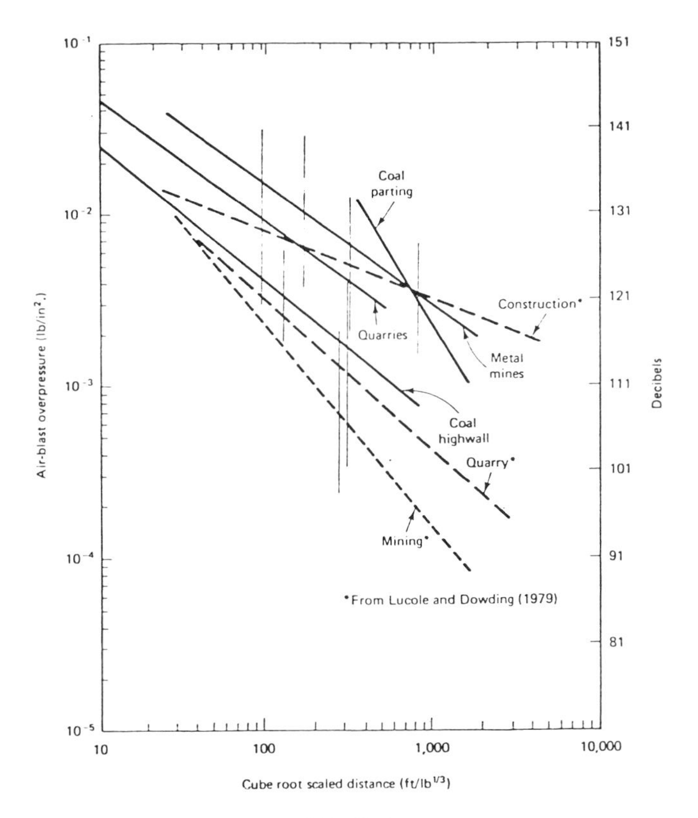

FIGURE 7
Attenuation of Peak Pressures as a Function of Type of Industry

Propagation of air blasts differs most greatly from that of ground vibrations because of the large influence of the weather. Wind and temperature inversions can cause the air blasts to focus in local zones of higher than expected pressures.

Various effects of the wind have been reported and should be added to the average relations presented in Figures 6 and 7. Kamperman's (1975) study of near-field air blasts produced by quarrying showed that air-blast pressures were 10 to 15 dB higher downwind than with no wind. Wiss and Linehan's study (1978) of air blasts produced by surface coal mining showed that in moderate winds the typical 7.7-dB reduction for each doubling of distance is reduced by

$$7.7 - 1.6V_{mph}\cos\theta \quad \text{dB} \tag{2}$$

where V_{mph} is the wind velocity in miles per hour and θ is the angle between the line connecting the blast and transducer and the wind direction.

An air temperature inversion occurs when the normal decrease in temperature with altitude is reversed because of the presence of a warmer upper layer. The warmer layer causes the sound pressure wave to be refracted back to the ground. Schomer (1973) has shown that for propagation distances of 2 to 40 miles, inversions produce zones of intensification of up to 3 times average values, with an average increase of 1.8 times (5.1 dB).

AIR-BLAST-INDUCED CRACKING

Air blasts generated by typical blasting operations are not likely to crack structures or their windows. However, the audible noise portion of the air blast and wall vibrations may startle people. Unusual air blasts are likely to crack windows before walls and often result from detonations with (1) large quantities of detonating cord at the surface, (2) little or improper stemming, (3) open fractures radiating from blast holes, or (4) above ground explosions such as for building demolition. In urban settings, the required blasting mats in combination with soil overburden, can be employed to reduce air blasts.

Since the sensitivity of windows to air-blast-induced cracking is a function of their size, thickness, and mounting, cracking is likely to be probabilistic in nature. Studies of claims from large accidental explosions (Reed et al., 1963; Redpath, 1976) have indicated the cracking probability of a single large pane (6 m^2 or 64 ft^2) as 1×10^{-4} at 130 dB and a single small pane (0.3 m or 3.5 ft^2) as 1×10^{-4} at 141 dB.

Observation of the response of specific windows yields an expected wide range of results that are excellently summarized by Siskind et al. (1980a). For instance, Taylor and Scott (1975) described a case where 30 greenhouses were subjected to a line charge that produced air-blast pressures of 140 dB and broke 239 out of 35,000 panes. Sutherland (1968) showed that precracked (3 ft x 3 ft x 1/8 in.) windows would not fail until sonic boom pressures exceeded 145 dB.

Thunder rarely breaks windows, yet as shown in Figure 4, the associated air blast produces at least 133 dB measured on a linear scale. This record was captured because the thunder produced ground motion of 0.09 ips and triggered a continuously sensing, computerized blast vibration monitoring system (DVI, 1984).

HUMAN RESPONSE TO AIR BLAST

Human sensitivity to air blast can be inferred from studies of sonic booms (Borsky, 1965; Kryter et al., 1968). As shown in Figure 4, air pressure time histories of sonic booms are similar to those produced by blasting. Human response to sonic booms, in terms of % of people annoyed, has been plotted in Figure 8 (Siskind, 1980a). As shown by the dotted line, when 50% of the impulses exceed 125 dB and 95% are less than 130 dB, 5% of the affected will be very annoyed. These values are based upon linear measurement systems. The thresholds should be much lower if C weighted instruments are employed, as indicated by the measurement differences shown in Figure 6.

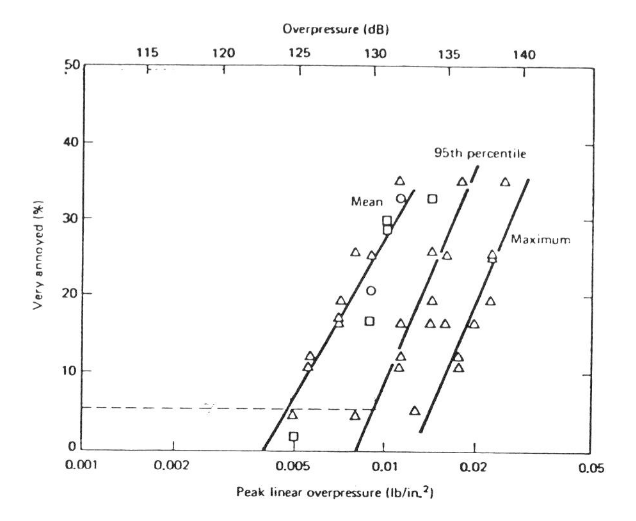

FIGURE 8

Percentage of Very Annoyed People after Sonic Boom Exposure where "Mean" Indicates 50% of Boom Pressures Were Larger Than the Plotted Value

REFERENCES

Borsky, P. N. (1965), "Community Reactions to Sonic Booms in the Oklahoma Area," Final Report for Aerospace Medical Research Laboratory, Wright-Patterson Air Force Base, AMRL-TR-65-37, Vol. II.

Dowding, C. H. (1984), Blast Vibration Monitoring and Control, Prentice Hall, Englewood Cliffs, NJ

DVI (1981), Private Communication from S. W. Lucole, chief engineer, Digital Vibration Inc., Northfield, IL

Kamperman, G. (1975), "Quarry Blast Noise Study for the Institute of Environmental Quality," Illinois Institute of Environmental Quality, Springfield, IL, 37 pp.

Kamperman, G., and Nicholson, M. A. (1970), "The Transfer Function of Quarry Blast Noise and Vibration into Typical Residential Structures," Environmental Protection Agency Report EPA 55/9-77-351.

Kryter, K. D., Johnson, P. J., and Young, J. R. (1968), "Psychological Experiments on Sonic Booms Conducted at Edwards Air Force Base," National Sonic Boom Evaluation Office, Washington, D. C. 84 pp.

Redpath, B. B. (1976), "A Review of Airblast Induced Window Damage," Proceedings of the Second Conference on Explosives and Blasting Techniques, Society of Explosives Engineers, Montville, Ohio, pp. 200-207.

Reed, J. W., Dade, B. J., Minor, J. E., and Deltart, R. C. (1963), "Evaluation of Window Pane Damage Intensity in San Antonio Resulting from Madina Faculty Explosion on November 13, 1963," Annals of the New York Academy of Science, New York.

Schomer, P. D. (1973), "Predicting Community Response to Blast Noise," U. S. Army Corps of Engineers Civil Engineering Research Laboratory, Technical Report E-17, 96 pp.

Siskind, D. E., Stachura, V. J., and Raddiffe, K. S. (1976), "Noise and Vibrations in Residential Structures from Quarry Production Blasting," U. S. Bureau of Mines, Report of Investigations 8168.

Siskind, D. E. Strahura, V. J., Stagg, M. S. and Kopp, J. W. (1980a), "Structure Response and Damage Produced by Airblast from Surface Mining," U. S. Bureau of Mines, Report of Investigations 8485.

Sutherland, L. C. (1968), "Sonic and Vibration Environments for Ground Facilities, A Design Manual," Wyle Laboratories, Report WR 682, prepared udner NASA Contract NAS8-11257, Huntsville, Ala., pp. 12-1 to 12-42.

Taylor, A. C., and Scott, D. S. (1975), "Quarry Blast Atmospheric Wave (Concussion) Response to Structures and Human Annoyance," Report of Ontario Ministry of the Environment, Ontario, Canada, 28 pp.

Wiss, J. F., and Linehan, P. W. (1978), Control of Vibration and Blast Noise from Surface Coal Mining, Research Report for the U. S. Bureau of Mines, Contract 10255022.

Vehicle Sound Power Measurement
Using the Acoustic Intensity Method

Richard D. Miech and Kenneth L. Thompson

ABSTRACT

This paper describes the use of the two microphone Acoustic Intensity Method of obtaining accurate measurement of over-all sound power of large wheel loaders. This system is in operation, within a factory environment, at the Caterpillar Aurora Plant. Its primary purpose is to assure that specified vehicle sound limits are being maintained. The system also provides diagnostic opportunities if units fail the validation test.

THE DEVELOPMENT OF THE ACOUSTIC INTENSITY METHOD of measuring sound power has provided an accurate means of measuring over-all vehicle sound level. Caterpillar Research and Development has been instrumental in integrating this method for use within an active factory environment. Although the individual components and hardware were available, the integration of these into a useable factory system required development and validation. The result has been a system that will test various size vehicles in a minimum amount of area, in a timely manner and at moderate cost. The need for more costly facilities such as an anechoic chamber or reverberant room for routine testing is negated. Space allocated for the Acoustic Intensity test area can be used for other productive purposes when not being used for sound testing. This may not be the case with the special facilities. Being an indoor system, it is not affected by weather elements such as rain, wind, snow or temperature.

This system of measuring and recording data also provides a good statistical base to aid Engineering in making design changes for sound level reduction, and the ability to measure the effects. Diagnostic capabilities provided by this system are most helpful to Manufacturing, Quality Control and Engineering in determining the causes for units that may fail to meet the sound limit requirements.

This paper was presented at the Society of Automotive Engineers 35th Annual Earthmoving Conference held in Peoria, Illinois, April 11, 1984. Copies of this paper are available by ordering #840768, SAE Order Department, Dept.792, 400 Commonwealth Drive, Warrendale, PA 15096 USA. Phone Orders 412/776-4970.

Richard D. Miech and Kenneth L. Thompson, Caterpillar Tractor Co., P.O.Box 348, Aurora, IL 60507

Cost Assessment of Construction Noise Control

Frederick M. Kessler, Ph.D.*

Abstract

The results of a series of studies sponsored by the Corps of Engineers' Construction Engineering Research Laboratory have been provided noise control engineers with a method to estimate off-site construction noise levels and the cost of reducing these noise levels to acceptable levels. From these studies, methods of selecting noise control methods for construction equipment and processes have been developed and presented in matrix format. Research into the cost of equipment noise control provides data which can be used to extrapolate costs to today's dollars. To test these findings, noise control at U. S. Army base construction sites were analyzed. Noise control for one of the noisiest equipment at a construction site, a pile driver, was designed and estimates for costs were made. Noise control effectiveness and costs were evaluated at a demonstration project.

Introduction

Construction noise control has always been considered as an expensive undesired expenditure by most developers and construction contractors. Replacement of old, noisier equipment by new, quieter equipment may cost a contractor more to purchase or rent. Contractors are also concerned that installation of noise control equipment may interfere with the construction process, slowing progress and increasing labor and operating expenses.

The objective of the Corps of Engineers studies was to quantify the cost of noise control and to compare these costs with the benefits which are derived (1),(2),(3),(4). A cost-benefit methodology was developed for use by cost estimators for what often is the noisiest phase of construction activity: Excavation and Foundation.

The cost-benefit methodology first requires an estimate of the construction site contribution to off-site noise and an evaluation of the noise control required, if any. A construction noise estimation procedure has been developed and improved which utilizes the construction equipment

----

*President, FMK Technology, Inc., P.O. Box 168, Bound Brook, New Jersey, 08805

maximum A-Weighted sound levels* and the percentage of time the equipment is in its noisiest operating mode, to compute the average sound level emitted from the site.

The off-site construction related sound levels are compared with noise impact criteria, as suggested by federal, state, or local agencies to determine the amount of construction site noise reduction which is required. Each construction phase requires different noise control techniques as a function of the amount of noise reduction which is required. Data showing the type of noise control for each construction phase to accomplish the desired noise reduction have been prepared. The cost for the selected noise control method can be readily estimated and a cost-benefit assessment can be made.

To demonstrate the methodology, a flood control project was selected by the U.S. Army Corps of Engineers where considerable pile driving was required (5). Effective pile driver noise control was designed and tested in the field. Costs associated with each noise control effort were determined.

The noise control estimating procedures, noise reduction selection procedures, and the pile driver demonstration project are discussed in the following sections of this paper.

Construction Site Noise Model

Estimates of construction site sound levels at each noise sensitive land use area should be made for each major phase of construction activity. The construction site sound level is estimated by adding construction equipment average sound levels *and extrapolating these levels to a location of interest. This estimating procedure provides the off-site construction average sound level for the time period of interest, i.e., one hour, one shift, one day, etc. Some state and local jurisdictions have adopted criteria for enforcement purposes which can be compared with the construction site's estimated off-site noise level for an early evaluation of potential regulatory problems.

At a construction site, a large number of machines are usually at work at numerous locations, each performing its functions using different work cycles. Equipment noise sources may be stationary, or nearly stationary, or mobile. During their work cycle their noise is heard by an off-site receptor to rise and fall in magnitude. Estimating off-site noise using only the maximum sound emitted by the construction equipment is excessively conservative and will over estimate the site average sound level by a large

---

* See nomenclature section for definition of Average Sound Level. All sound levels used in this paper are A-Weighted unless noted otherwise.

amount. The following methodology considers this variation in equipment noise emissions.

The average sound level of an individual equipment source is computed from the maximum sound level measured when the equipment is doing useful work. This information is supplied by the manufacturer and has been published in a number of government reports, including those listed in the References section of this paper.

The maximum sound level is modified by a Usage Factor which is defined as the percentage of time the equipment typically operates in its noisiest mode. Usage Factors have been measured and reported. A knowledge of the construction methodology can be used to estimate Usage Factors when needed information has not been found in the literature. The Usage Factor must relate to the period of construction for which the site average sound level is being estimated, i.e., one hour, one shift, daytime (15 hours), nighttime, etc.

To obtain the Average Sound Level for the construction site, the average for each item of equipment must be added using the following relationship:

$$L_{eq} = \sum_{i=1}^{N} n_i \times UF_i \times 10^{(L_{Ai}/10)} \quad ,\text{dB} \tag{1}$$

where L_{eq} is the Average A-Weighted Sound Level in dB for the construction site

n_i is the number of units of the ith type of construction equipment.

UF_i is the Usage Factor for the ith type of construction equipment.

L_{Ai} is the maximum A-Weighted sound level for the ith type of construction equipment.

The construction equipment is assumed to be operating at some hypothetical center of activity. If the construction site is very large, more than one computation is made by assuming that equipment is grouped at more than one center of construction activity.

This model produces estimates which are reasonably accurate. The model has been tested at construction sites where different phases of construction had been taking place. The following table provides an indication of the accuracy of the model.

	Boundary Line Average Sound Levels, dB	
	Measurements	Model
Interior Finishing	66	66
Roofing	60	61
Erection of Houses (a)	69	71
Erection of Houses (b)	63	63
Grading	70	68
Fill and Grade	70	75
Clearing	73	72

Noise Impact Criteria

The U.S. Army Construction Engineering Research Laboratory has suggested boundary line noise level for different phases of construction. They are:

	Average Sound Level at Boundary, dB
Demolition	83
Excavation	78
Foundation (no pile driving)	78
Foundation (with pile driving)	83
Framing	71

Note: these limits are designed to encourage the contractor to keep predominant work areas at least 100 Ft. (30 m) from noise sensitive areas.

For long term construction projects, such as a nuclear power plant or a large public works project, the federal Department of Housing and Urban Development criterion (6) of a Day-Night Average Sound Level of 65 dB can be used.

State and local regulations might apply and should be researched. Many state and local codes use maximum sound levels not average sound levels as the measure of noise impact. The above model can be modified to estimate an absolute worst-case situation by setting the Usage Factor to one (1.00).

Feasible Noise Control Methods

Feasible noise control methods have been studied for different phases of construction activity(3),(4). The phases for which noise control scenarios were developed are; Grading, Backfill/Excavation, Trenching, and Foundation. Additional research into the noise control methods for the other phases of construction is needed. Grading activities have been further subdivided into three sub-categories; Road Grading, Site Grading, and Street Grading. Trenching has been subdivided into Ditching and Filling the Trench. The Foundation Phase has been subdivided into three sub-phases;

Sheet Piles, Concrete Preparation, and Concrete Footing.

Other sub-phases are used in construction but since they have not been studied by this investigator, they have not been presented in this paper.

For each construction phase (and sub-phase), a construction scenario was created for the noise emission estimates and noise reduction evaluations. For example, for the backfill phase - Four scraper/haulers, working simultaneously, move fill from a pile and distribute it (backfill) on the site. Crawler tractors, crawler or wheeled loaders,or heavy graders are sometimes used to perform similar functions. A water truck is used to contain the fugitive dust.

After each scenario was conceived, feasible noise control methods were designed to reduce site noise by a) less than 3 dB, b) 3-6 dB, c) 6-10 dB, and d) 10-15 dB.

The most commonly used noise control methods for construction sites fall into three broad categories:

1. Barriers;
2. Equipment Modifications; and
3. Equipment Substitutions.

In addition, proper scheduling of the noisiest activities can be used as an effective noise control means. Since some construction activity regulations relate to average sound levels for a given period, it may be advisable to distribute the noise producing work so that the criterion is not exceeded. If the situation warrants, quieter equipment can be used together with the noisy equipment, raising the average sound level only slightly, but producing periods of quiet when neighbors can get relief. The cost benefit for scheduling has not been studied and should be considered on a case-by-case basis.

The following table presents a matrix of feasible noise control methods. The letter-number symbols used in the table are described below:

Construction Phases	Symbol
Road Grading	G1
Site Grading	G2
Street Grading	G3
Rough Backfill	E1
Site Backfill	E2
Ditching	T1
Filling the Trench	T2
Sheet Piles	F1
Concrete Preparation	F2
Concrete Footings	F3

Noise Control Methods	Symbol
Plywood Barriers	B1
Stockpiles	B2
Earth Berms	B3
Cooling System	M1
High Performance Mufflers plus M1	M2
Engine Enclosure, Intake Silencer, Vibration Isolation, plus M2	M3
Equipment Substitutions	S1
Use of Offsite Facilities	S2

Feasible Noise Control Methods

Noise Reduction	Construction Phases								
	Grading			Backfill/ Excavation	Trench		Foundation		
	G1	G2	G3	E1, E2	T1	T2	F1	F2	F3
Less than 3 dB	M1 B1 S1	M1 B1 B3 S1	M1 B1	B1 B3 M1 S1	B1 B3 M1	B1 B3 M1 S1	M1 B1	B1 B2 B3 S2	M1 B1
3-6 dB	B1 B1 + M1 M2 B1 + S1	B1 B3 B1 + M1 M2 B1 + S1 B3 + M1 B2 + S1	B1 B1 + M1	B1 B3 M2 B1 + S1 B3 + S1 B1 + M1	B1 B3 M2 B1 + M1 M1 + B3	B1 B3 M2 B1 + S1 B3 + S1 B1 + M1	M2 B1 M1 + B1	B1 B2 B3 S2	M2 B1 M1 + B1
6-10 dB	B1 B1 + M1 B1 + M2 M3 B1 + S1	B1 B3 B1 + M1 B1 + M2 M3 S1 + B1 B3 + M1 B3 + M2 B3 + S1	B1 B1 + M1 B1 + M2 M3	B1 B3 M3 B1 + S1 B3 + S1 B3 + M1 B3 + M2 B1 + M1 B1 + M2	B1 M3 B3 + M1 B3 + M2 B1 + M1 B1 + M2	B1 B3 M3 B1 + S1 B3 + S1 B3 + M1 B3 + M2 B1 + M1 B1 + M2	M3 B1 M1 + B1 M2 + B1	B1 B2 B3 S2	M3 B1 M1 + B1 M2 + B1
10-15 dB	B1 + M2 B1 + S1 B1 + M3 B1 + M1	B1 + M2 B1 + S1 B3 + M3 B1 + M3 B1 + M1	B1 + M2 B1 + M3 B1 + M1	B1 + M1 B1 + M2 B1 + M3 B1 + S1 B3 + M3 B3 + S1	B1 + M1 B1 + M2 B1 + M3 B3 + M3	B1 + M1 B1 + M3 B1 + S1 B1 + M2 B3 + M3	M1 + B1 M2 + B1 M3 + B1	B1 B2 B3 S2	M1 + B1 M2 + B1 M3 + B1

Note the matrix element with a heavy border. This boxed example indicates the noise control techniques which can be used to reduce site grading noise by 3-6 dB. A plywood barrier or earth berm can accomplish 3-6 dB of reduction depending upon where the barrier is placed. A plywood barrier plus equipment cooling system modification can achieve the desired noise control as will high performance mufflers plus cooling system modification without barriers.

The cost of achieving the feasible noise control presented in the above matrix can be analyzed and estimated. Equipment modification is most easily determined through discussions with the equipment manufacturers. For example, in 1977, the percentage increase in construction phase cost for equipment modifications are shown below:

Equipment Modification	Construction Phase Scenario						
	G1	G2	G3	E1	E2	T1	T2
M1	.29	.39	.55	.35	.25	.42	.33
M2	.54	.75	1.00	.68	.49	.75	.60
M3	1.40	2.00	2.80	1.80	1.30	2.10	1.70

The maximum increase in construction costs is less than 3% due to the purchase of noise control components for construction equipment for the scenarios which were studied.

Costs for the other noise control methods can be assessed on a case-by-case basis. The cost for a barrier, for example, depends upon the height and length of the barrier used.

Pile Driver Demonstration Project

Could a contractor for a large construction project include noise control in the project, assess its costs, and accomplish this noise control effectively for approximately the costs he had estimated earlier? A flood control project in Waterloo, Iowa where an extensive amount of pile driving was required was selected in an attempt to obtain answers to the above question. Specific objectives of the demonstration project were:

1. to demonstrate that pile driver noise can be reduced through retrofit controls incorporated at the construction site;
2. to show that the bid document can be effectively utilized to effect noise control;
3. to affirm that a construction contractor is capable of a realistic noise control bid estimate which will closely approximate his actual costs; and
4. to give evidence that a construction contractor can effectively minimize construction site noise without extensive instructions or outside technical assistance.

Draft specifications for the quieting of pile drivers

at the Waterloo, Iowa flood control site were developed and incorporated into the Corps of Engineers Civil Works bid documents as a separate bid item. These specifications were such that all of the five contractors who bid on the project were able to provide realistic bids (a detailed cost analysis was conducted by the project team prior to the issuance of the bid documents). None of the contractors found it necessary to require additional information from the project team although an offer for assistance was contained in the bid specifications.

Seventeen (17) piles were used in the demonstration project as part of 36 piles used in the construction of a gravity outlet structure constructed at the intersection of the Virden Creek with the Cedar River in the city of Waterloo. The piles were 18 in. round pipes which were later filled with concrete.

A Vulcan single-acting pile driver (hammer) size 010 was selected by the contractor as the standard unit. A standard Vulcan muffler, "Decelflo", an off-the-shelf unit, and an enclosure (see Figure 1) was designed by the contractor to fit the Vulcan hammer. The Vulcan hammer provides a rated striking energy per blow of 32,500 Ft. lb. at 50 blows per minute. Its rated striking velocity is 14.51 Ft./Sec. The unit weighs 10,000 lbs.

The noise control retrofit methods which were tested were:

a) Enclosure - The enclosure was basically an acoustically treated metal container which is attached to the hammer to shield the impact area when the hammer/anvil strikes the pile (see Figures 1 and 2). The enclosure was made of 18-gauge steel lined on the inside with acoustic foam to reduce the build-up of sound within the enclosure due to reverberation. The enclosure was 11 Ft. 10 in. long. A 4 Ft. vinyl skirt was attached to the bottom of the enclosure so as to completely enclose the impact area. Plastic windows were installed in the enclosure and the skirt so that the hammer action could be observed by the operators.

b) Muffler - An off-the-shelf muffler with an A-Weighted sound level reduction of 15 dB was selected by the contractor to reduce the noise caused by the sudden release of exhaust air. The muffler was specifically designed by Vulcan for use on the exhaust ports of their air and steam driven pile drivers. The cost of the unit was $3,457.00 in 1978.

c) Pile Damping - A water based damping compound was applied in an uncontrained form in rings about 1 in. thick at 8 Ft. intervals on the piles. Since the application of the damping treatment is a costly, time consuming process, only a few piles received this treatment.

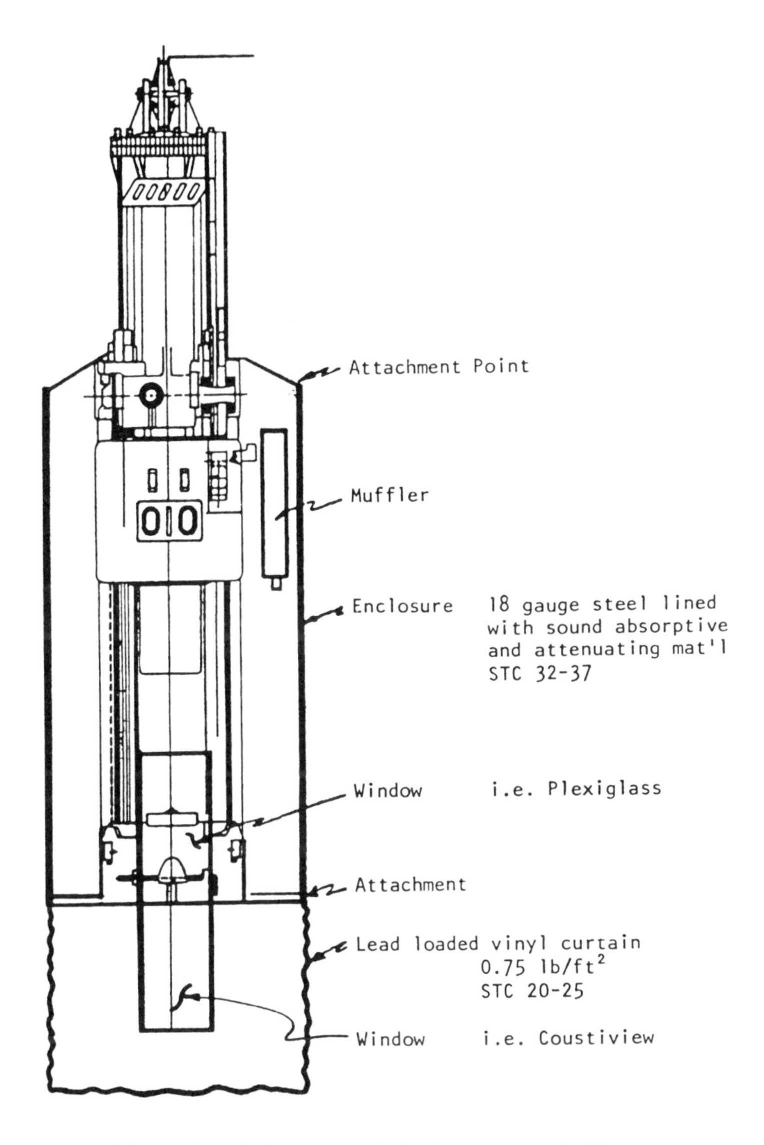

Figure 1. Schematics of Enclosure and Muffler

Figure 2. Photograph of Enclosure

d) Impact Cushion - An impact cushion (pad), constructed of 4 to 12 in. of alternating layers of 1/4 in. sheet aluminum and 1/2 in. phenolic material has been used to reduce the hammer impact noise. For this demonstration project, 1 in. thick elastomeric material was substituted for the phenolic layers.

e) Alternative Pile Drivers - Two alternative pile drivers were tested. A McKiernan-Terry, MKT V-20, vibratory pile driver was tested at the site. A Bolt unit based on a Chelminski design was tested at the Bolt Associates plant in Norwalk, Connecticut.

Sound level measurements were made from a boat moored in the river 264 Ft. and on the shore 100 Ft. from the gravity outlet, respectively. Sound emissions were recorded using a Bruel and Kjar Model 4921 outdoor microphone system and a Nagra DJ magnetic tape recorder. The recorded data were analyzed using a GenRad Model 1921 Real-Time-Analyzer controlled by a Digital Equipment Corporation PDP 8/e computer.

The measured maximum sound level (L_1, the sound level exceeded 1% of the time, was used to represent the maximum sound level) and the average sound level for the cases tested are presented below. These results represent the averaging of a number of pile driver runs.

Condition	A-Weighted Sound Level at 264 Ft.	
	Maximum	Average
Unsilenced	87	80.7
Enclosure/muffler/damping/pad	76	70.4
Enclosure, no muffler	80	73.2
Vibratory pile driver	70	68.1
Bolt (distance scaled)	80	70.8

The maximum and average sound level for two cases where piles were driven with a muffled and enclosed pile driver with a) pile damping but no pad and b) a pad but no pile damping were:

a) Silenced with pile damping	76	70.0
b) Silenced with pad	81	72.7

From the above data, we can see that the contractor did an effective job (about 10 dB) of reducing the average sound emitted during pile driving operations. The effect of the use of these noise control efforts on the capital and operating costs are discussed in the following paragraphs.

Capital costs are reflected in the amortized or rental charge on a weekly operating basis. Included with the weekly operating costs are the labor costs associated with the installation of the damping treatment and the pile cushion (pads). The operating costs are reflected by the productivity data obtained from the contractors logs for each pile which was driven. The following table contains the one-week pile driving costs.

	Cost $/Week	% Increase Over Standard
Standard unit	11,597	-
Enclosure, no muffler	11,626	0.3
Enclosure/muffler	11,646	0.4
Enclosure/muffler/damping	16,196	39.7
Enclosure/muffler/damping/pads	16,430	41.8
Enclosure/muffler/pads	11,880	2.4
Vibratory unit	12,603	8.7

Since the muffler and enclosure can be used for many years, their capital cost per week does not increase the weekly pile driver cost by 1%. But, use of the damping treatment due to the labor intensive effort required for installation of the damping increases the weekly pile driving cost by almost 40%. The vibratory pile driver is most effective in reducing the noise emissions but its use increases the weekly operating cost by about 9%. For some cases where the noise reduction is critical, this cost increase is necessary. But, vibratory pile drivers do not operate effectively in all types of ground conditions and a standard unit, such as a Vulcan, may have to be selected.

The time to drive the pile to the desired depth is an important measure of effect on performance of the noise control measures. Set up time is also important. It was noted that the set-up time improved for the duration of the demonstration project. At the beginning some difficulty was experienced with the set-up of the pile driver with the enclosure. After a number of piles were driven the set-up period decreased significantly. A sufficient number of piles were not driven by the pile driver with the enclosure for a reliable estimate of set-up time to be determined. The average driving time comparisons are listed below:

	Time/Pile Minutes	Depth Feet	Blow-Rate Blows/Min.
Standard Unit	7.25	19.9	64.6
Enclosure/Muffler	-	30.5	-
Enclosure	12.0	28.0	58.8
Enclosure/Muffler/Damping	11.0	22.5	40.6
Enclosure/Muffler/Damping/Pads	11.6	28.5	56.6
Enclosure/Muffler/Pads	9.5	26.0	77.2
Vibratory Unit	18.3	26.2	-

It is difficult from a review of the above table to quantify the effect on performance due to noise control. The Blow-Rate is a function of the source air pressure and the subsurface conditions. The Blow-Rate affects the time it takes to drive the pile to its desired depth. It appears from the above data that the enclosure and other noise control measures minimally affect the pile driver performance. The vibratory pile driver is less effective in driving piles the desired depth. Subsurface conditions can render the vibratory pile driver inneffective and a standard pile driver must be brought to the site to complete the job of driving the piles.

Conclusions

Noise control measures can be applied to pile driving in a cost-effective manner to reduce the objectionable off-site noise by about 10 dB.

The request-for-bid document containing noise limit specifications or noise control requirements is an effective means for achieving construction site noise reduction.

A construction contractor has the resources to design or select the appropriate noise control measures to meet the requirements contained in the bid documents.

The cost for the noise control can be estimated reasonably accurately. The effect upon operating costs must be determined on a case-by-case basis. For pile driving, the increase in weekly costs is not significant (less than 1%) and performance effects are also not significant.

Nomenclature

Average Sound Level is the equivalent steady A-weighted sound level containing the same sound energy as the time-varying sound. Technically, the average sound level in decibels (dB) is the integral of the A-weighted mean square sound pressure devided by the time period taken for the average.

A-weighted Sound Level is the sound level measured after the ambient sound is filtered by an A-weighting network prescribed by the American National Standards Institute (ANSI) S1.4-1971, Specifications for Sound Level Meters.

Day-Night Average Sound Level is the 24-hour average sound level, in decibels, dB, obtained after addition of 10 dB to nighttime (10:00 PM to 7:00 AM) sound levels.

Exceedance Sound Level (also known as Percentile Sound Level) is the sound level exceeded x-percent of the time during the period of observation. For example:

L_1 is the A-weighted sound level exceeded 1% of time during the observation period.

Sound Pressure Level is devined as 20 times the logarithm (base 10) of the ratio of the root mean square sound pressure to a reference standard sound pressure of 20 microPascals. It is defined as

$$20 \, \mathrm{Log}_{10} \, (P_{rms} / P_0)$$

where P_0 is the standard reference sound pressure - 20 microPascals.

References

1. Schomer, P.D. and Homans, B., "Construction Noise: Specification, Control, Measurement, and Mitigation," CERL Interim Report E-53, April, 1975

2. Schomer, P.D., et al, "Cost Effectiveness of Alternative Noise Reduction Methods for Construction of Family Housing," CERL Interim Report N-3, July, 1976

3. Kessler, F.M., et al, "Construction-Site Noise Control - Cost-Benefit Estimating Procedures," CERL Technical Report N-36, January, 1978

4. Kessler, F.M., et al, "Construction-Site Noise Control Cost-Benefit Estimating Technical Background," CERL Technical Report N-37, January, 1978

5. Kessler, F.M. and Schomer, P.D., "Noise Control: Pile Driver Demonstration Project, Waterloo, Iowa, CERL Technical Report N-111, July, 1981

6. Department of Housing and Urban Development, Environmental Criteria and Standards, Title 24, Code of Federal Regulations, Part 51, July, 1979

Highway Noise:
FHWA Requirements and Barrier Design

Grant S. Anderson*

Abstract

Since 1973, the Federal Highway Administration (FHWA) has required noise studies for nearly all proposed roadways and roadway improvements that receive federal aid. This paper begins with an overview of these FHWA requirements today: requirements for the study process itself, for the computation methods, and for noise-barrier design. The overview highlights the broad nature of the FHWA requirements, and the particular care needed to avoid regulatory and technical pitfalls.

Next, this paper details the design process for recent noise barriers along the Long Island Expressway in New York City. Included on the design team were two civil/structural engineers, an acoustical engineer, the State's project manager, the regional FHWA representative, and staff knowledgeable in safety, maintenance, and aesthetics. Described here are details of the interaction among these disciplines in the noise-barrier design.

Overview of FHWA Requirements

In response to the Federal-Aid Highway Act of 1970, in February of 1973 the FHWA promulgated its PPM 90-2: Procedures for Abatement of Highway Traffic Noise. Since then, these procedures have undergone periodic revision. The current FHWA regulation is part of the Federal-Aid Highway Program Manual and is commonly known as FHPM 7-7-3: Procedures for Abatement of Highway Traffic Noise and Construction Noise.[5]

FHPM 7-7-3 applies to nearly all Federal-aid highway projects involving new construction or significant improvements. For these, it (1) stipulates noise-study procedures for both operational and construction noise, (2) describes noise-abatement measures that can be federally funded and that must be incorporated where feasible and reasonable, and (3) sets noise-level criteria for such noise-abatement measures. (In addition, it requires communication with local officials, which is not discussed further in this paper.)

The requirements of FHPM 7-7-3 can be grouped as follows:

o <u>Noise Measurements</u>: Required is the determination of existing noise levels within the highway corridor. These existing noise levels provide a baseline against which "relative" impact is

*Supervisory Consultant, Bolt Beranek and Newman Inc., 10 Moulton Street, Cambridge MA 02238

judged: How much will the proposed highway increase noise? Although not strictly stipulated, this baseline is almost always measured, rather than predicted mathematically.

- Noise Predictions: Required are predictions of project traffic noise for design-year conditions.

- Impact Assessment: Impact is assessed in two ways: first in a "relative" way by comparing design-year predictions with the existing-noise baseline, and second in an "absolute" way by comparing predictions with the FHPM's Noise Abatement Criteria, which depend only upon land use.

- Noise Abatement: Required is the incorporation of noise-abatement measures into the project's plans and specifications, wherever impact is identified and such measures are both feasible and reasonable.

- Construction Noise: Required is (1) identification of potentially affected land use, (2) identification of abatement measures whose benefits outweigh adverse effects and costs, and (3) inclusion of such measures in the plans and specifications.

All these requirements are stated broadly by FHWA. FHPM 7-7-3 leaves a great many details to individual project engineers, as do all such federal regulations. In the field of noise control, however, no state design standards exist to provide the missing details. Confusion often results, and the pitfalls are many. The next several sections of this paper contain particularly troublesome pitfalls, with some suggestions for overcoming them.

Potential Noise-measurement Pitfalls

Outdoor noise measurement is an inexact practice, full of difficulties not found with indoor measurements. Do not consider indoor experience sufficient. Take a course in outdoor measurements, or thoroughly read one of the standard texts.[3, for example]

Outdoor noise often fluctuates wildly, as noise sources come and go. Include only those sources that are typical for that measurement site; exclude the others. Choose measurement instrumentation that allows such exclusion of typical noise sources.

The temptation always exists to calibrate the prediction mathematics against the measurement results, and to modify the prediction mathematics where disagreement is found. The FHWA does not allow this. If a simple check between prediction and measurement is made, recognize that most of the bias and scatter may be in the measurements, not in the prediction mathematics.

A wet pavement greatly increases tire noise. Do not measure when pavements are wet unless you qualify the results.

Wind direction and speed can greatly affect noise levels from sources as close as 500 feet (152 meters). When measuring at this distance or greater, recognize that the results may not be typical of other wind conditions. It is best to measure during wind typical of that site.

Potential Noise-prediction Pitfalls

FHPM 7-7-3 requires use of a prediction model that is "generally consistent" with the FHWA published model.[2] The phrase "generally consistent" is not clear. Use the actual FHWA model in all cases. It is available in three levels of complexity: (1) a hand-calculation method backed by programmable calculator, (2) a "level-1" computer program, called SNAP, for uniformly cross-sectioned highways, and (3) a "level-2" computer program, called OPTIMA, for highways with complex geometrics. If some other model is used, obtain official sanction prior to its use. Some states have their own prediction models, already sanctioned by FHWA.

The computer program SNAP is tempting to use because of its simplicity. However, it significantly overestimates noise-barrier performance adjacent to upgrades, and introduces many simplifying assumptions that do not always apply. Avoid SNAP; use the "level-2" program STAMINA almost exclusively.

In using STAMINA, horizontal curves must be approximated by straight-line segments. The user's manual provides no guidance on how precise one must be here. Approximate curves precisely wherever receptors lie close to the roadway.

Concerning noise emissions from single vehicles, the FHWA requires use of either its national-average emissions or emissions measured by its procedures. [2,4] The problem here lies with heavy-truck emissions when these trucks are: (1) on off-ramps, (2) accelerating to mainline speed after leaving on-ramps, (3) anywhere near traffic signals, and (4) full-throttle in anticipation of an upgrade or after just cresting a hill. The throttle habits of truckers do not depend solely upon roadway grade, as the FHWA model assumes. In all these circumstances, the national-average emissions are inaccurate and may result in over-estimates of the noise level. In such circumstances, measure your own heavy-truck emission levels and have them sanctioned by FHWA.

The FHWA model does not allow speeds less than 30 miles per hour (50 kilometers per hour). If such speeds are input, STAMINA automatically increases them to this minimum. This FHWA limitation is severe and unrealistic in urban areas and for ramps. It is best to bypass this limitation in the program and insert realistic vehicle emissions at these lower speeds.

None of the three FHWA methods works well for stop-and-go traffic in urban areas. The methods contain inappropriate vehicle emissions and ignore the complex pattern of noise reflections between the source and receptor. Currently no official model exists that is adequate in such urban areas. For projects in such areas, it is best to (1) measure vehicle emissions specific to the project and to (2) develop a propagation model specific to the urban terrain. This second task is nearly always impossible within time and budget constraints. Be prepared for low-accuracy predictions. Think about modeling the propagation with an acoustical scale model.[1]

When sound propagates for 200 feet (60 meters) or more across grass, the sound level is reduced for receptors within ten feet (3 meters) or so of the ground. To approximate this reduction, the FHWA model incorporates its "ground-effect" mathematics. This mathematics is very approximate, however, and can lead to strange anomalies. The strangest is that it predicts a discontinuous noise level at ten feet (3 meters) above the ground. Be aware of this and be ready to rationalize the anomaly in the technical appendix.

Potential Noise-impact Pitfalls

Impact is assessed in two ways: first in a "relative" way by comparing design-year predictions with the existing-noise baseline, and second in an "absolute" way by comparing predictions with the FHPM's Noise Abatement Criteria, which depend only upon land use. Although the temptation is there, the noise impact of a "build" alternatives should not be assessed by comparison to the "no-build" alternative. Such an assessment is against FHWA policy. An analogy here with safety is appropriate. Concerning safety, it is not enough that the "build" alternative be as safe as the "no-build." Required is that the "build" alternative meet current safety standards, independent of how unsafe the existing conditions are. The same is true for noise. Current noise standards must be met; it is not enough to just hold the noise to its existing value, or to its "no-build" value in the design year. Do not fall into this trap.

The "relative" assessment compares design-year predictions with the existing-noise baseline, to yield the noise increase caused by the project. FHPM 7-7-3 provides no numerical criterion for this noise increase. Most states have decided upon their own criterion, for both the descriptive wording and the abatement implications. Use state policy here.

The "absolute" assessment compares design-year predictions with the FHPM's Noise Abatement Criteria. For example, the outdoor criterion for residences is 67. (Units are purposely avoided here.) This '67' is not the onset of impact, however; impact occurs for noise levels lower than 67. In fact, FHWA's technical advisory group originally recommended criteria some 10 decibels lower. However, these lower criteria were judged not feasible, nationwide, by FHWA. Because noise impact exists well below the Noise Abatement Criteria, do not argue that impact is minimal because the Noise Abatement Criteria are exceeded by only one or two decibels.

Potential Noise-abatement Pitfalls

Abatement must be considered wherever traffic-noise impacts are identified: either "relative" or "absolute" impacts. Either is sufficient by itself; both are not needed. Do not assume that a minimal noise increase means no barriers are needed.

In considering an abatement option, its feasibility and reasonableness determine whether or not it should be incorporated into the project. FHPM 7-7-3 contains no guidelines on feasibility and reasonableness. Some less-official FHWA guidelines do exist, however.[6] They discuss common misapplications of FHPM 7-7-3. According to these guidelines:

- Do not eliminate a noise barrier just because its cost exceeds that of the protected properties.
- Do not eliminate a noise barrier for safety and maintenance reasons. In almost all instances, problems associated with safety or maintenance can be resolved during the design process.
- Do not set a maximum barrier height. Determine height by comprehensive acoustical design and/or the desires of the affected residents, not value engineering alone.
- Consultation with the highway neighbor on the reasonableness of barrier designs is good practice.

The FHWA's computer program OPTIMA allows cost-effective barrier design. Cost savings up to 40 percent are possible, without reduction in barrier effectiveness. Use this program; it merges with STAMINA, which computes all the acoustics.

The "absolute" assessment compares design-year predictions with the FHPM's Noise Abatement Criteria. These criteria are rough hewn. They do not distinguish between subtle differences in land use, or even between some gross differences. For example, the criterion for parks is the same as that for playgrounds. Even for the same "absolute" assessment, noise barriers may be more justified for some of these land uses than for others. Take this into account for critical land uses.

FHPM 7-7-3 requires consideration of a long list of noise-abatement options: traffic management, alignment changes, noise barriers, buffer zones, and noise insulation of public-use of nonprofit institutional structures. Consider them all and discuss them all in the environmental documents.

The greatest abatement emphasis has historically been on controlled-access roadways. Nevertheless, abatement is possible on uncontrolled-access roadways, too. Even noise barriers are possible there where frontage roads exist. Do not ignore uncontrolled-access roadways.

Often the predicted noise will exceed the Noise Abatement Criterion by only one or two decibels. In such situations, do not design a noise barrier to achieve only a one-or-two-decibel reduction. Such a reduction will not be perceptible. Instead, aim for a healthy reduction of 6-or-more decibels. Remember that noise impact occurs even when the Noise Abatement Criterion is exceeded by zero decibels.

The FHWA prediction model assumes an effective heavy-truck source height of eight feet (2.4 meters), even though the stacks of heavy trucks are often higher. Check that noise barriers interrupt the line-of-sight between a 12-foot (3.7 meters) stack and receptors. If not, it is wise to increase the barrier height.

Where an elevated structure passes over a project roadway or lies adjacent to it, noise can reflect from the underside of the structure and thereby pass over top of a noise barrier. This reflection can greatly reduce the noise barrier's performance. Be aware of such a possibility and calculate the reflected energy separately.

Multiple reflections in urban areas can easily reduce barrier effectiveness. Where such reflections may exist and where barrier performance is critical, consider acoustical scale modeling.

Often barriers must be built on both sides of a roadway, to shield neighbors on both sides. Such "parallel" barriers produce multiple reflections back and forth across the roadway. In turn, these multiple reflections compromise the barrier performance on both sides. This compromise is not incorporated into the FHWA prediction model. It does appear in the FHWA's Noise Barrier Design Handbook, however.[7] Incorporate the handbook's parallel-barrier nomograph into the predictions to determine the compromise. To avoid the compromise, either make the interior barrier walls acoustically absorptive or tilt the barriers back from one another to "spill" the reverberation towards the sky.

Durable outdoor acoustical absorption is difficult to engineer. Where such absorption is needed, investigate its physical properties thoroughly.

Gaps through or under noise barriers have far more effect than would seem reasonable. The sound energy that gets through such gaps can substantially compromise a barrier's performance. Avoid all gaps in barriers.

Often a barrier must incorporate an access gap for maintenance of the right-of-way. Where such a gap is necessary, provide it by overlapping two barrier segments, so that the sound must "bend" several times before heading towards the neighbors. A 15-to-20 foot (4.6-to-6.1 meter) overlap is generally sufficient.

Especially in urban areas, the project roadway might not be the predominant source of noise for many receptors. To compute barrier effectiveness, all other sources of noise must be taken into account, as well. Often these other sources preclude achieving significant noise reduction with a project-roadway barrier. The non-project noise sources, which are not quieted by the barrier, prevent significant net reduction. Take this into account in barrier design, and do not build barriers that are severely limited in this way; they are a complete waste of money.

When sound propagates for 200 feet (60 meters) or more across grass, the sound level is reduced for receptors within ten feet (3 meters) or so of the ground. After a barrier is built, however, the sound path is forced upwards and this "ground-effect" attenuation mostly disappears. As a result, the barrier's benefit is less than simple mathematics would predict. The benefit is reduced by the amount of the lost ground-effect attenuation. Be sure to take this loss into account; it is not explicitly calculated in the FHWA model.

Wind effects barrier performance. Downwind, performance is worsened; upwind it is improved. The FHWA model does not take this wind effect into account. If prevailing winds blow from source to receptor, design for a greater barrier performance than actually sought. Even better, compute the sound-ray's curvature and adjust the barrier results of the FHWA model.

If trees protrude above a barrier top, they may scatter sound into the shadow zone behind the barrier, thereby reducing the barrier's effectiveness. This effect is not included in the FHWA model. Avoid large masses of such trees, or measure their effect under similar geometric conditions.

Potential Construction-noise Pitfalls

FHPM 7-7-3 sets no criterion for construction noise. The temptation is to set a criterion equal to existing noise levels, to prevent noise increase during construction. Whether or not such a criterion is feasible will depend upon the "averaging time" used in the construction-noise analysis. For example, a phase-long averaging time might well make this criterion feasible. With this averaging time, construction scenarios are approximated over a full phase of construction, and then the phase-average noise level is computed. This phase-average computation averages the very loudest moments (when equipment is closest and operating) with the quietest moments, to result in one moderate value. Such a long-term averagd value does correlate best with annoyance. On the other hand, averaging over much shorter periods -- single hours, for example -- will produce a whole range of computed levels, from high to low. Most likely, it will not prove feasible to reduce the highest of these down to existing-noise levels. Be sure to take the averaging time into account in setting construction-noise criteria. Otherwise, the criteria may be infeasible to attain.

Standard texts often catalog the costs of various noise-abatement items for construction noise. In reality, such costs vary greatly from project to project. A great part of such costs consist of delay-time costs where noise abatement slows construction. Be wary of published abatement costs; estimate such costs separately for each project.

A Sample Barrier Design Process

Detailed noise-barrier design was recently undertaken along a four-mile portion of the Long Island Expressway in New York City. The design proceeded in three steps: (1) initial computations, (2) design by committee, and (3) modifications by neighbors. Described here are some details of the interaction among the various disciplines involved in this detailed barrier design -- details concerning barrier heights and positions.

Initial Computations

First, traffic projections and highway plans were assembled. Next, several acoustical decisions were made: (1) Noise emissions were chosen for slow-speed vehicles, including slow-speed buses. (2) Special on-ramp coding was decided, to account for the continued acceleration of heavy trucks, even after leaving an on-ramp. (3) Prototype predictions were made for several times of day, to decide when the worst noise impact occurred. (4) A preliminary decision was made concerning which windows would be considered "open" and which "closed," for assessment of indoor noise impact.

Next, the roadway was coded into the STAMINA computer program, in manageable portions. Receptors were decided and also coded — first floors only.

Next, STAMINA was run, to obtain no-barrier noise levels, separately for project and non-project roadways. Where project noise did not exceed the applicable Noise Abatement Criterion, then that receptor was dropped. Also dropped were receptors where noise from frontage roads and local streets precluded significant barrier benefits. Then, based upon these two STAMINA runs, a design goal was set for each remaining receptor. This design goal depended upon the receptor's total noise, its Noise Abatement Criterion, and its balance between project and non-project noise.

Next, STAMINA was used to design a system of first-attempt barriers, aimed at meeting the design goal at each receptor. These first-attempt barriers were then optimized with the program OPTIMA, and results were reported to the Barrier Design Team.

Design by Committee

The Barrier Design Team consisted of two civil/structural engineers, an acoustical engineer, the State's project manager, the regional FHWA representative, and staff knowledgeable in safety, maintenance, and aesthetics. This team met for several full-day working meetings to decide upon preliminary barrier design. In all, three additional barrier iterations were undertaken, each consisting of (1) discussion about the non-acoustical constraints that had to be met, and then (2) a redesign and reoptimization of all the component barriers.

The first-iteration discussions resulted in the following barrier changes:

- Rethinking about which windows should be considered "open" and which "closed" for assessment of indoor noise impact.
- Inclusion of train noise from a crossing rail line, as it effects the maximum benefit achievable by the barriers in that vicinity.
- Reduction of barrier heights for barriers on structure.
- Decision on minimum safety clearance at on- and off-ramps.
- Addition of several more receptors.

- Addition of some second-floor receptors, which would then receive as much emphasis in the design as would the first-floor receptors.

- Modification of the horizontal position of some of the barriers, to better fit the cross-sectional geometry and for safety purposes.

- Elimination of some barrier portions that destroyed essential pedestrian access.

- Inclusion of a second barrier design in one neighborhood -- a design shifted a large distance horizontally from the original design.

After these non-acoustical decisions were reached, the barriers were redesigned and reoptimized with STAMINA/OPTIMA, while retaining several height options for each neighborhood.

The second-iteration discussions resulted in the following additional barrier changes:

- Elimination of several height options.

- Elimination of some barrier elements, while making adjacent elements higher to maintain barrier performance -- all for aesthetics.

- Slight modifications to some receptor locations.

- Addition of several height-constrained designs, where computer optimization recommended extreme heights.

After these additional non-acoustical decisions were reached, the barriers were again redesigned and reoptimized with STAMINA/OPTIMA. This final set of computer computations included upper floors as well, to document the vertical extent of barrier protection. Barrier performance was tabulated as the difference between the noise level with and without barriers: the so-call Barrier Insertion Loss.

Modifications by Neighbors

In each barrier neighborhood, meetings were held with the neighborhood Community Planning Board, to describe the barrier-design process and to request community input. Subsequent meetings were held with neighbors who directly abut the project. At these neighborhood meetings, two types of demonstrations were held:

- Visual demonstrations of the appearance of the proposed barriers.

- Acoustical demonstrations of the noise reduction that the barriers would achieve.

Neighborhood input was then requested concerning barrier height. Some neighborhoods wished the heights reduced, in which case the barriers were redesigned and reoptimized -- the third iteration. Often the heights could not be reduced significantly while retaining a worthwhile benefit. This was explained to the abutting neighbors. And last, heights were finalized with the neighbors, and final barrier benefits were recomputed.

Summary

This paper included an overview of FHWA noise-study requirements, highlighting the potential pitfalls where particular care is needed. Included were pitfalls concerning noise measurements, noise predictions, impact assessment, noise abatement, and construction noise.

Next, this paper detailed the design process for recent noise barriers along the Long Island Expressway in New York City. Discussion included details of the initial computations, the barrier design by committee, and barrier modifications by neighbors. Details were limited to those concerning barrier heights and positions.

APPENDIX: REFERENCES

1. Anderson, G.S., "Acoustical Scale Modeling of Roadway Traffic Noise: A Literature Review," Bolt Beranek and Newman Report No. 3630, Cambridge MA, April 1969. (Study undertaken for the FHWA Office of Research and Development.)

2. Barry, T.M. and J.A. Reagan, "FHWA Highway Traffic Noise Prediction Model," Report FHWA-RD-77-108, U.S. Federal Highway Administration, Washington DC, December 1978.

3. Beranek, L.L., Noise and Vibration Control, Field Measurements: Equipment and Techniques, McGraw-Hill, 1971.

4. Bowlby, W., "Sound Procedures for Measuring Highway Noise: Final Report," Report No. FHWA-DP-45-1R, Federal Highway Administration, Washington DC, August 1981.

5. Federal Highway Administration, "Procedures for Abatement of Highway Traffic Noise and Construction Noise," FHPM 7-7-3, Transmittal 348, Washington DC, 9 August 1982. Also commonly referenced as 23 CFR 772.

6. Reagan, J.A., "A National Field Review of the Highway Traffic Noise Impact Identification and Mitigation Decisionmaking Processes," Memorandum from the Chief, Noise and Air Analysis Division, Office of Environmental Policy, Federal Highway Administration, HEV-30, 8 June 1982.

7. Simpson, M.A., "Noise Barrier Design Handbook," Report No. FHWA-RD-76-58, Federal Highway Administration, Washington DC, February 1976.

STANDARDIZING NOISE BARRIER FIELD MEASUREMENTS

by

William Bowlby*, M. ASCE

Abstract

American National Standards Working Group S12-6 was organized in 1979 to draft a standard for the acoustical measurement of outdoor noise barriers. This paper describes the current version of the draft standard and outlines the remaining work.

Introduction

American National Standards (ANSI Standards) in the areas of physical acoustics, bioacoustics, mechanical shock and vibration, and noise are published by the Acoustical Society of America (ASA). Accredited Standards Committee S12 within ASA deals with noise. In 1979, Working Group S1-75 (now S12-6) was organized to draft a standard to measure the acoustical performance of outdoor noise barriers [1]. The group includes experts from government (Federal, State and Canadian), academia and consulting.

Typically, such groups work strictly on a volunteer basis. However, because the Federal Highway Administration (FHWA) was very interested in such a standard for traffic noise barriers, it provided research funding to assist in the production. By means of a contract to USDOT Transportation Systems Center (TSC) with a subcontract to ASA, assistance has been provided for writing, and travel to meetings, work sessions and field evaluations. In addition, funds were allocated to TSC for field testing. The final product of the FHWA contract will be a document that is ready for the ANSI review process on its way to becoming an approved ANSI standard.

The draft standard adopts insertion loss--the difference in levels before and after noise barrier installation--as the basis for determining effectiveness of a barrier. Currently, the purpose of the draft standard is given as [10]: "This standard provides methods for determining the insertion loss of outdoor noise barriers at receiver locations of interest under conditions of interest. In addition, this standard presents requirements for the documentation of the procedures and results to permit interpretation and independent evaluation of the results."

---

*Research Assistant Professor of Civil Engineering, Vanderbilt University, Box 96-B, Nashville, TN 37235

The draft standard specifies that the resultant insertion losses are valid only for the given measurement conditions. It does not present methods for comparing performance of barriers at different sites or for generalizing barrier performance for other conditions. While one might be able to make comparisons between sites or extrapolations to other conditions, such efforts would be beyond the scope of the proposed standard. Further, any such comparisons or extrapolations would have to be justified and documented.

Determining insertion loss provided by outdoor noise barriers is usually difficult. Difficulties arise from the lack of measured levels at study sites prior to barrier installation, or from the inability to accurately estimate these "before installation" levels. Lack of a standard method may lead to incorrect or poorly documented results, with independent evaluation of the techniques and results not being possible.

In an effort to overcome these problems for one of the most common applications of noise barriers, the Federal Highway Administration published a field measurement manual in 1981 [2]. The draft standard includes many of the ideas contained in the FHWA manual, but is generalized to cover other types of outdoor noise barrier applications.

Scope

The draft standard does not define standardized receiver locations. Measurements may be made at receivers of interest under conditions of interest, within certain prescribed limits. For example, if desired, the methods could be used to determine the insertion loss of a construction noise barrier at a receiver position between two buildings; this insertion loss could be in terms of an hourly average sound level for when the wind is blowing at 2 km/h from the site to the microphone. Likewise, the standard could be used to measure the sound exposure level insertion loss of individual truck passages for an open field when winds are calm.

However, use of the standard would be with the constraint that the measured insertion loss would only apply to the stated conditions. It could not be generalized to represent other conditions without documented justification. It is for this latter reason, in part, that the draft standard calls for careful documentation of the measurement conditions.

Use of the A-weighted sound exposure level, A-weighted one-hour average sound level, or octave band sound pressure level is recommended. However, the draft standard does not preclude use of other noise descriptors or measurement durations appropriate to the purposes of the user. Determination of octave band insertion loss is intended for more research-oriented applications than for standard barrier performance evaluation. For octave band measurements, use of a more regimented array of receiver positions is suggested in addition to use of receiver positions of interest.

The draft standard covers the use of sources naturally occuring at a site, controlled natural sources, and controlled artificial sources. Also, for cases where direct measurements cannot be made prior to barrier installation the draft standard presents two methods for indirectly determining these BEFORE levels: (1) by measurement at a second site that has been judged as equivalent to the study site prior to barrier installation, or (2) by prediction. Equivalence refers to source characteristics, receiver locations, terrain, and atmospheric and ground conditions. Because of the broad scope of the draft standard, it only presents principles, not procedures, for judging source and terrain equivalence, as well as for choosing a prediction model.

The draft standard also presents a method for determination of a "lower bound" to the insertion loss. Such a value may be all that can be determined when clear measurement of the source level behind the barrier is prevented because of high background noise or low levels when the source is operating. In many applications, the reporting of a barrier insertion loss of "equal to or greater than X dB" is sufficient for the purposes of the user.

Finally, the standard presents a method for accounting for the bias and random errors associated with the insertion loss determination. Reliable and repeatable results may be expected from use of the proposed standard when the distances between all receivers and their closest sources (or closest point of approach for mobile sources) are within 100 metres. While use beyond this range is not precluded, the experimental error will likely increase substantially.

Methods

Three methods for insertion loss determination are presented in the draft standard. The recommended method is the "direct measured" method. An "indirect measured" method using measured BEFORE levels on an equivalent site is also given, as well as an "indirect predicted" method using predicted BEFORE levels. Choice of a method may be dictated by several factors. These include the objectives of the study, the ability to make measurements prior to barrier installations, and the ability to achieve source, terrain, ground and atmospheric equivalence between the BEFORE and AFTER situations.

The "direct measured" method may only be used if the barrier has not yet been installed or can be removed for the BEFORE measurements. In this method, insertion losses are determined from actual measured levels at the site both before and after barrier installation at the receiver positions of interest. For a valid determination, however, the levels must be measured with equivalent sources, ground conditions, and atmospheric conditions. For example, the source cannot change in output level, spectral content, or location without justified and documented adjustments to the measured levels. If the natural source is expected to change significantly from BEFORE to AFTER, the draft standard suggests use of a controlled natural or artificial source.

If the barrier has been installed prior to any actual BEFORE measurements and it cannot be readily removed, the draft standard directs the user to simulate the BEFORE condition by measuring at a site that is equivalent to the study site, except for the presence of the barrier. Equivalence will be discussed in subsequent sections of this paper.

If direct or equivalent BEFORE measurements cannot be made by either of the first two methods, the draft standard presents a method using substitute BEFORE predictions. When predictions are used, however, errors inherent in the prediction method will further decrease the precision of the results. Moreover, the draft standard requires that the the resulting insertion loss be referred to as "partially measured".

The draft standard does not prescribe use of particular prediction methods. However, it does require that the prediction method take into account the type of ground between source and receiver. Also, documentation must exist on the field validation or calibration of the prediction method for the relevant test conditions, and must be referenced in the test report. The quantitative error from this validation is then incorporated into the assessment of experimental errors called for in the draft standard. Finally, source, receiver and reference microphone positions must be modeled as precisely as possible relative to the ground within the constraints of the chosen prediction method.

Sources

The draft standard specifies that three types of sources may be used for the noise measurements. The actual source for which the barrier is designed is normally the best source. However, if an investigation of the natural sources indicates a lack of equivalence between BEFORE and AFTER cases, the draft standard suggests use of a controlled natural source. For example, if truck flow rates at a traffic noise barrier site are expected to change significantly, it may be useful to select one or several test trucks to be typical sources, and then use them for controlled BEFORE and AFTER tests.

When BEFORE/AFTER natural source equivalence cannot be established, or when preliminary testing is desired, the draft standard recommends use of a controlled artificial source. However, a great deal of technical expertise is required for use of artificial sources to ensure relevance of the tests to the natural source(s) being simulated. Furthermore, the draft standard requires that this relevance be established and documented in the test report.

Two ways of ensuring source equivalence for BEFORE and AFTER measurements are described in the draft standard. The aim is to establish that the source level, spectrum, directivity and position (particularly the effective height above the ground) is the same for both BEFORE and AFTER tests. The first way is the monitoring of operational parameters that affect the acoustic output. The second way is the monitoring at a reference position of the actual acoustic output in the direction of the receivers.

No single set of operational parameters is applicable to all sources to which the draft standard applies. Appropriate parameters may be defined by trade associations, standards organizations or government agencies. For example, for purposes of the draft standard, accepted parameters for traffic noise are average operating speeds, and separate hourly flow rates for each vehicle class. As another example, operational parameters for construction machinery may be developed at the site by observing the source operational cycle or by interviews of the operators. In any case, the draft standard calls for careful documentation of source parameters, so that test conditions can be duplicated for the AFTER measurements.

The draft standard also specifies principles for the location of a reference microphone to monitor for source equivalence. It states that the position 1.5 m (5 ft) directly above the top of the barrier is the normal reference position. However, where the distance from source to barrier is too small to neglect near-field effects, alternative locations are suggested.

Terrain Equivalence

To permit valid comparisons of BEFORE and AFTER sound level measurements for insertion loss determination, the draft standard requires that the equivalence of terrain, ground, and atmospheric conditions between the BEFORE and AFTER cases be determined and documented. The process of evaluating terrain equivalence is complicated [5]. When sound is propagated over a reflective plane, there are regions where sound waves combine or cancel out each other. The magnitude of these effects can be quite large, especially for narrow band and pure tone point sources propagating over a flat ground plane. However, the effects are usually reduced as sources become larger or more distributed, as frequency band width is enlarged, and as terrain becomes more undulating in character.

The draft standard gives principles for the assessment of equivalence. First, one must determine that the same major noise components of the source would be seen at the receiver position for the two sites (presuming absence of the barrier for the AFTER site). Second, similar reflection patterns off the ground between the receiver and the major noise components of the source must exist at each site. Also, similar reflective paths must exist from vertical surfaces near the receiver position.

The difficulty of terrain equivalence calculations is strongly dependent on source characteristics. The Working Group intends to publish separately a suggested method for terrain equivalence evaluation for highway traffic noise, which may be useful as a model for developing methods for other sources.

Ground Equivalence

The draft standard says that the equivalence of the ground at two sites may be determined by measurement of the acoustic impedance of the ground. However, it notes that while several methods exist to measure this property [3, 6, 9], there are no standard procedures. Therefore, as a minimum, the user is required to "tightly categorize" the ground and select an equivalent site of a similar category. Examples of such categories are: paved; long grass over loose sand; and short grass over packed clay. Also, the draft standard directs the user to avoid extreme changes in the surface water content, as well as frozen or snow-covered ground because of its variability.

The vegetation at a site is usually important only insofar as it keeps the ground loose and porous. Normally, there is only appreciable scattering by leaves on bushes and trees for frequencies above two to three kilohertz. In densely forested areas, however, there may be some reverberation due to reflection from trunks and branches. Thus, equivalent sites should have a tree density that is roughly comparable, while bushes and occasional trees may be ignored.

Atmospheric Equivalence

The major atmospheric phenomena that affect sound propagation are wind, temperature and cloud cover. The draft standard states that atmospheric conditions should be chosen to be relevant to the activities of people affected by the noise or to the objectives of the study.

The draft standard proposes different classes of wind conditions and stipulates that the acoustical measurements before and after barrier installation be made in the same wind propagation class. That class is determined by monitoring wind speed and direction, determining average values, and computing the vector component of wind velocity in the direction of sound propagation. This vector component puts the acoustical measurements in one of the five classes defined in the draft standard. In all cases, measurements are restricted to when the average wind velocity is below 6 m/sec (13 mi/hr).

Temperature and Cloud Cover

The draft standard also calls for consideration of ambient air temperature and cloud cover. Cloud cover affects vertical temperature gradients and, hence, sound propagation. Three basic cloud cover classes are defined in the draft standard. The user is directed to make the BEFORE and AFTER acoustical measurements under the same class of cloud cover and when the average temperatures are within 25° F of each other.

For octave band measurements, the draft standard gives additional guidance on meteorological monitoring, especially for research into atmospheric effects on barrier performance. In such cases, determination of the vertical profile of the temperature is recommended.

Background Noise Levels

Requirements for consideration of background noise are also contained in the draft standard. High background noise levels may mask measurement of the actual level of the source, thus preventing determination of the full insertion loss potential of the barrier. In other cases, it may not be possible to reduce the source level to measure the background level.

For a true measure of unmasked insertion loss, the draft standard requires that the background level be 10 or more dB below the level of the source of interest. If the background level is within 10 dB, the draft standard presents a method to determine an adjusted insertion loss. If the actual background level cannot be measured, the standard describes how to determine a lower bound to insertion loss.

Instrumentation

The draft standard calls for use of sound level meters or measurement systems meeting the Type I requirements per ANSI S1.4-1983 [7]. Use of Type II equipment is not precluded, but will increase experimental error. The draft standard recommends use of at least two measuring systems, to permit simultaneous measurements at a receiver position and a reference position. Use of integrating sound level meters is also recommended if the chosen descriptor for the study is average sound level or sound exposure level.

For octave band sound pressure level measurements, the draft standard directs the user to choose octave bands based on the characteristics of the source being studied. Typically, eight such bands, corresponding to geometric mean frequencies of 63, 125, 250, 500, 1000, 2000, 4000, and 8000 Hz, would be used.

The draft standard also calls for some means of characterizing the source to document its equivalence for the BEFORE and AFTER measurements. The means will depend on the type of source. For example, for traffic noise, vehicle speed could be measured by using radar instruments, while vehicle emission levels could be measured using a sound level meter, and vehicle classification counts could be made using counter boards.

Measurement Duration or Sample Size

Because of the broad scope of the proposed standard, a specific measurement duration or sample size is not required. Choice of a duration by the user should consider several factors, including the temporal variation of the noise, the noise descriptor being measured, and the range in sound level or sound pressure level.

For guidance on judgment of the temporal nature of the noise, the draft standard refers users to ANSI S1.13-1971(R1976) [8]. ANSI S1.13 defines the following categories of noise:

1. steady (e.g., air-conditioner)
2. non-steady
   a. fluctuating (e.g., traffic)
   b. intermittent (e.g., train passage)
   c. impulsive
      (1) isolated bursts (e.g., pile driver)
      (2) quasi-steady (e.g., riveting, pneumatic hammer)

Measurement repetitions at each receiver position are required to assess the repeatability of the results. The number of repetitions to be made at a receiver position will affect the random error that must be computed and reported. The draft standard presents suggested measurement durations and sample sizes based on the temporal nature of the noise and the range in sound level fluctuations.

Measurement Procedures

The basic measurement procedure first calls for the background noise level to be sampled. After measuring the background noise, the user is then directed to sample the level with the source in operation at each microphone, including the reference microphone. These steps are then repeated until all desired source-receiver pairs have been studied. The background level should be sampled with the source shut off or removed from the site. If that is not possible, the draft standard suggests a method for determining an upper limit to the background level. In such cases, however, a true unmasked insertion loss cannot be determined; instead only a lower bound to insertion loss can be determined.

For the "indirect predicted" method, the user is directed to make the on-site measurements for the AFTER case paying special attention to collection of all site and source data required for use of the chosen prediction model. For example, at a traffic noise barrier site the user may want to sample vehicle emission levels to ensure that the values in the model are accurate for the particular traffic being measured. Then, the user would predict the BEFORE levels at the reference position and at each receiver position using the measured source operational parameters and the observed terrain and ground conditions.

Computations

The draft standard gives detailed steps to determine an insertion loss (or its lower bound) and the experimental error for each source-receiver pair, as permitted by background level criteria. For octave band applications, these values would be computed and recorded separately for each octave band studied.

The calculation process first involves adjustments to the measured source levels for calibration level shift and for background noise bias. The latter adjustment covers situations when background noise levels are measured in absence of the source, as well as when the source is in operation.

For each measurement repetition the insertion loss at each receiver is determined by subtracting the difference in adjusted reference and receiver levels for the BEFORE case from the difference in adjusted reference and receiver levels for the AFTER case:

$$IL_i = (L_{AR} - L_{Ai}) - (L_{BR} - L_{Bi})$$

where

IL_i is the insertion loss at the ith receiver for the given source
L_{Bi} and L_{Ai} are, respectively, the BEFORE and AFTER adjusted levels at the ith receiver from the given source.

L_{BR} and L_{AR} are, respectively, the BEFORE and AFTER adjusted reference levels for the given source.

A mean insertion loss, $\overline{IL}_i$, or its lower bound, is then computed for each receiver.

An important part of the reported insertion loss is the experimental error. This error includes both random and bias errors. Random errors include instrumentation error, variation in levels within a given wind class, and variation in levels within a given cloud cover class. In addition, there may be unknown biases, such as:

1. calibrator bias within its stated accuracy range;
2. error inherent in prediction model; and
3. equivalent site bias (for indirect measured method).

Random errors are considered in the calculation process by the repetition of the measurements for each source-receiver pair and the resultant computation of a variance for the measurements. Bias errors are accounted for by approximation of variance for each bias error and the addition of these variances to the random error variance.

Reporting Requirements

Finally, the draft standard gives detailed requirements for reporting of results. Documentation must be sufficiently complete to permit independent repetition of the experiment, evaluation of the results, and evaluation of the validity of any assumptions made in the study.

Remaining Work

The current draft of the proposed standard is now under field evaluation. A test plan was prepared for FHWA and is slowly being implemented. The field evaluations are aimed at testing the clarity and completeness of the writing as well as the validity of the measurement and calculation procedures.

So far, one full field evaluation using natural sources has been finished by the Florida Department of Transportation [4]. Other State highway agencies, including California, Iowa, Pennsylvania and New Jersey, have offered paper reviews, or provided barrier insertion loss data collected previously by their own methods. Finally, FHWA plans to conduct a thorough field evaluation when its traffic noise barrier research site is completed on the Dulles International Airport property in northern Virginia.

Once the field tests are completed, the draft standard will be finalized and submitted to FHWA to complete the contract. It will then also begin the formal ASA/ANSI review/balloting process on its way toward planned adoption as an approved American National Standard.

APPENDIX I - REFERENCES

1. Anderson, G.S., "Acoustical Measurements of Outdoor Noise Barriers: Progress on Standardization," Proceedings of the 1980 International Conference on Noise Control Engineering, Poughkeepsie, NY, 1980.

2. Bowlby, W. (ed.), Sound Procedures for Measuring Highway Noise: Final Report, Report No. FHWA-DP-45-1R, Federal Highway Administration, Arlington, VA, 1981.

3. Embleton, T.F.W., J.E. Piercy and G.A. Daigle, "Effective Flow Resistivity of Ground Surfaces Determined by Acoustical Measurements," Journal of the Acoustical Society of America, Vol. 74, pp. 1239-44 (1983).

4. Lindeman, W., Noise Barrier Insertion Loss Research Study, St. Petersburg, Florida, Florida Department of Transportation, Tallahassee, FL, 1984.

5. Rickley, E.J., An Investigation of Site Effects on Roadside Measurements of Truck Noise, Report No. DOT-TSC-OST-76-6, USDOT Transportation Systems Center, Cambridge, MA, 1976.

6. Zuckerwar, A.J., "Acoustic Ground Impedance Meter," Journal of the Acoustical Society of America, Vol. 73, pp. 2180-6 (1983).

7. ----, American National Standards Specifications for Sound Level Meters, ANSI S1.4-1983, Acoustical Society of America, New York.

8. ----, American National Standards Specifications for the Measurement of Sound Pressure Levels, ANSI S1.13-1971 (R1976), New York.

9. ----, ASTM Standard C384-54, "Impedance and Absorption of Acoustical Materials by the Tube Method," American Society for Testing and Materials, Philadelphia, 1958.

10. ----, Proposed American National Standard Methods for Determination of Insertion Loss of Outdoor Noise Barriers, ANSI S12.8-198X, Acoustical Society of America, New York, unapproved, unpublished.

The Noise Mitigation Plan for Stapleton Airport

Andrew S. Harris*

ABSTRACT

In March 1984, seven communities in the vicinity of Denver's Stapleton International Airport completed a year-long study of methods to reduce the adverse noise effects from aircraft operations at Stapleton. The study resulted in a Noise Mitigation Plan that could significantly improve the noise environment. The author reports on the process the communities followed to develop the Plan, describes the 13 noise mitigation measures it incorporates, and reviews progress toward implementation in the time since the study ended.

INTRODUCTION

In this paper I shall focus on the special issues associated with noise mitigation at Denver's Stapleton International Airport (Stapleton) and the recent Noise Mitigation Study (Study). I shall, however, place the Stapleton Study in the general context of airport noise control.

Noise problems occur at airports when the noise from aircraft operations become loud enough in off-airport communities to interrupt activities or cause annoyance. Large areas immediately adjacent to airports are frequently devoted residential uses, uses that are incompatible with the existing levels of aircraft noise. Although airports were often planned initially in undeveloped areas, two kinds of growth have occurred which result in conflicts between the airports and their neighbors. Airports often expanded their facilities, increased the numbers of operations or began to serve noisier aircraft than when they opened -- or grew in all of these ways. At the same time, communities around the airports allowed residential devlopment in the previously undeveloped land. In addition to these basic conflicts, the off-airport land typically falls within several political subdivisions.

The environment at Stapleton includes all of these elements. What made the Stapleton situation different, and particularly interesting to us, was that all of the major political subdivisions around Stapleton decided to work together on the problem. The City and County of Denver, owner of Stapleton, and six other communities around the airport sponsored the Study and participated in it.

*President, Harris Miller Miller & Hanson Inc., 429 Marrett Road, Lexington, MA 02173

In the balance of this paper I shall review the Study, how it was conducted and how it was posible to give a primarily lay group a working understanding of the complexities of noise abatement.

STUDY HISTORY

Each of the sponsoring communities had representatives on the Executive Committee, the body that provided overall guidance for the Study. A larger Noise Study Committee included not only representatives from neighborhoods of the sponsor communities but also representatives from the airlines, airline pilots, the Federal Aviation Administration, and other government agencies. The Executive Committee selected the firm of Harris Miller Miller & Hanson Inc. as the Study Consultant to work with the Committees. The Study Consultant worked with the Noise Study Committee to insure that the Study addressed all relevant concerns and, conversely, to make certain that committee members and their constituents understood each step of the Study.

STUDY GOAL

The Executive Committee adopted a goal for the Study:

> The primary goal of this study is to develop a series of practical, effective noise mitigation measures (a "noise mitigation plan") that will eliminate or minimize the present and future adverse effects from aircraft operations at [Stapleton].

STUDY METHOD

To accomplish this goal;, the Study had six basic tasks:

- o Identify current noise impacts;
- o Quantify current noise impacts;
- o Determine effects of present noise mitigation policies;
- o Select additional noise mitigation measures;
- o Determine effects of additional mitigation measures, and
- o Prepare the Noise Mitigation Plan.

Each of these tasks had three parts. The first part consisted of work by the Study Consultant. Usually, this was research. The second part was a Working Paper prepared by the Study Consultant describing the particular task and the results of the research. The members of the Noise Study Committee received a copy of the Working Paper for a task to allow them to prepare for the third part, a meeting to discuss the results of the task and prepare for the next step.

Task 1 - Identify Current Noise Impacts

During the first task, the Executive Committee and the Study Consultant asked area residents to describe their concerns about existing noise problems. In general, people felt airport noise interfered with conversation, radio and TV listening, might have adverse health effects, and degraded the use of their yards. They thought noise from takeoffs was the main problem. But, residents believed that noise had increased over the past years and they also expressed concern about aircraft deviations from flight corridors and lack of interest by authorities.

Task 2 - Quantify Current Noise Impacts

During the second task, the Study Consultant recommended the method for describing the noise and developed descriptions of the current noise environment. Airport-produced community noise is made up of a series of separate events. Between events, the environment may be relatively quiet. The primary method of describing aircraft noiser is the day-night average sound level, Ldn. Ldn reflects, in a single number, the amount of sound energy experienced during an average 24-hour day. This Study uses Ldn to describe the overall effects of airport operations.

In general, Ldn correlates well with community reaction to intrusive noise; the higher the Ldn value, the greater is the number of people who will be highly annoyed by the noise. An Ldn value of 65 is generally regarded as the threshold of unacceptable community noise; above this level, residential land use is inappropriate unless the residential buildings contain special sound isolating features, such as double-glazed windows and air conditioning that permit closed windows year-round.

Noise contours depict the pattern of exposure around an airport. They are not exact. For example, we can identify the neighborhoods exposed to a particular noise level. But, we cannot say which lots a particular noise contour crosses. Nonetheless, contours provide useful information about noise levels in areas around airports and they allow us to evaluate the effects of noise mitigation efforts.

The Study Consultant quantified noise exposure by determining (both measuring and computing) noise levels produced by current aircraft operations. The Committees examined effects of current mitigation policies, then they evaluated additional mitigation measures.

The Study Consultant faced an interesting challenge in working with Ldn. To use Ldn successfully in this Study, it was necessary to be sure that all members of the committees could understand the basis of Ldn and also understand how it could be used to evaluate the effects of noise mitigation measures. This was an educational process. First, the committee members developed an appreciation of the relationship between the noise from single departures or arrivals and the Ldn for a day. The first examples were for days when only one set of runways was used for each kind of operation. (At Stapleton, the most common configuration is to have landings on runways 26L and 26R with departures on runways 35L and 35R.) (See Figure 1 for the runway layout.) Sets of Ldn contour were prepared for each of the major runway-use configurations. Then, information was provided showing the percentage of a year that each major configuration was used. Finally, the Ldn contours were prepared for the annual-average usage pattern. By combining their knowledge of the daily Ldn and the annual usage pattern, committee members could understand how an area of the community could be noisy on those days when aircraft were flying in the area yet still not have high levels of annual Ldn, because the aircraft were not in the area during very many days.

MITIGATION MEASURES: PRESENT AND FUTURE

Task 3 - Determine Effects of Present Mitigation Noise Mitigation Policies

Operating procedures that existied at Stapleton at the start of the Study included two measures that had significantly changed the noise exposure from what it would have been if the measures had not been in effect. First, departures to the west and to the south were purposely reduced in preference to departures to the north and to the east when weather permitted. This preferential runway use reduced noise exposure to the west and south, while slightly increasing it to the north over Rocky Mountain Arsenal. Second, departing aircraft could not turn prior to reaching an altitude of approximately 2,200 feet above the ground. This restriction meant that aircraft were higher and, hence, quieter when flying over communities beside the airport than they would have been if they were permitted to turn at lower altitudes.

Task 4 - Select Additional Noise Mitigation Measures

During the Study, the committees and the consultant reviewed more than 20 separate potential noise mitigation measures. These measures included all of the options listed in the overview table (Table 2) in Federal Aviation Regulation Part 150 as promulgated by the Federal Aviation Administration. A number of measures were rejected as inappropriate in view of Stapleton's role as a major hub airport. Rejected measures included curfews, limitations on types of aircraft using the airport and limitations on the numbers of aircraft using the airport.

Task 5 - Determine Effects of Additional Mitigation Measures

This task considered the costs and benefits of each of the proposed measures. It also considered how the measures could be implemented.

Task 6 - Prepare the Noise Mitigation Plan

After thorough evaluation of potential measures, the Executive Committee adopted a Noise Mitigation Plan consisting of 13 measures. The measures are in four groups: changes in the airport plan; changes in runway use or airspace use; changes in land use, and; noise program management. I next describe briefly the measures included in the Plan.

CHANGES IN THE AIRPORT PLAN

Noise Barriers at Airport Boundary - Tall, continuous barriers (walls or berms) or buildings along the south boundary of the airport would protect homes just south of the boundary from landing noise (reverse thrust) and start-of-takeoff noise on the southern east-west runway, 8R-26L.

Aircraft Runup Area with Noise Barrier - Aircraft runup noise causes some complaints from neighboring communities. A runup area inside a barrier (a wall or earth berm with a wall on it) would reduce this noise. To accommodate the full range of commercial jets, the barrier should be 35 feet high.

CHANGES IN RUNWAY USE OR AIRSPACE USE

Departure Route Changes - Departing aircraft may not turn before achieving an altitude of 2,200 feet above ground level, as mentioned above. Though this restriction has improved the noise environment around Stapleton, further reduction of noise levels is possible if turns are permitted only after achieving 3,200 feet altitude. The Committee chose this 3,200 foot altitude as representing a reasonable compromise between numbers of people who would experience lower noise levels and the costs incurred in flying the added distance before turning on course.

Refined Preferential Runway Use System - Present mitigation measures include an effort to have departing aircraft take off to the north (as mentioned above) and arriving aircraft land from the east. This procedure tends to put the noisy takeoffs over the less densely populated area north of the airport. Currently, about 85% of departures and 75% of arrivals conform to this preferred mode of operation. It is likely that increases to 90% of departures and 80% of arrivals are possible. This would produce further improvements in the noise environment. The Plan includes efforts to increase daytime use of this pattern. In addition, it recommends that aircraft take off toward the north and land toward the south during nighttime hours.

The combination of departure route changes and refinement of the preferential runway use system could provide significant reductions in exposure for areas to the immediate west, south and southeast of the airport. The reductions in exposure would occur for one of the following reasons: (1) the aircraft would be higher and quieter; (2) there would be fewer flights near a particular area; or, (3) a combination of the previous two conditions.

Changed Helicopter Routes and Altitudes - Though helicopters probably contribute minimally to the overall noise environment, their noise is so identifiable that many people find it objectionable. The noise effects of helicopters would be reduced by: 1) increasing use of flight corridors along interstate highway routes; 2) increasing the altitude of helicopter flights.

CHANGES IN LAND USE

Building Code Provisions Related to Outside Noise - Building codes of jurisdictions surrounding Stapleton should include provisions that require a building to provide adequate isolation from outdoor noise. Residential buildings constructed in areas where outdoor levels exceed Ldn 65 would provide enough sound isolation so that resultant indoor levels are no louder than they would be if the outdoor noise did not exceed Ldn 65.

Compatible Use Zoning - Jurisdictions can prevent future noise problems by zoning so that land uses will be compatible with the long-term noise environment. If departure routes and runway use are altered so that high noise levels are moved to sparsely populated areas, the jurisdictions would not allow unrestricted residential development in those areas.

Sound Insulation of Existing Buildings - The intrusion of aircraft noise into buildings can be reduced by modifying windows and other openings. Modifications would assure that indoor noise from external sources is no higher than it would be if the exterior noise level did not exceed Ldn 65. Buildings exposed to long-term levels in excess of Ldn 65 would be candidates for treatment.

Acquisition of Property or Easements - Acquisition of property and conversion of existing developments is one way of eliminating incompatibility. Acquisition would be considered for areas exposed to levels exceeding Ldn 75. Acquisition feasibility is dependent upon consideration of resulting community disruption and upon cost of acquisition and relocation.

Noise Notices - Zoning maps and subdivision plats should include information about the noise environment where noise levels exceed Ldn 65. All jurisdictions would incorporate noise contours and explanations into their planning documents.

NOISE PROGRAM MANAGEMENT

Monitoring the Noise Mitigation Plan - All of the jurisdictions should monitor the progress and effects of implementation of the mitigation measures. The Plan describes methods for insuring that the progress of implementation is tracked. Methods include record keeping, "before and after" noise measurements, and periodic reports.

Continued Community Participation - All groups who are concerned with issues of airport-related noise must keep aware of each other's concerns. Community participation throughout the Noise Mitigation Study has produced increased understanding of the difficulties and complexity of airport noise control. If the Noise Mitigation Plan is to be efficiently and effectively implemented, community participation must continue. The plan proposes that the existing Executive Committee and Noise Study Committee form the core group of participants.

Upgrade Noise Complaint System - Presently, Denver treats noise complaints just as any other complaint received by the airport. An upgraded complaint system could make the interaction between the complainant and the airport useful to both parties. There should be a concentrated effort to document every complaint and complainants should received informed answers to their questions.

IMPLEMENTATION OF THE PLAN

Although the Study concluded with publication of the Final Report and Noise Mitigation Plan (1), the real work of noise mitigation just began at that time. Implementation of the 13 measures requires continued effort. The Executive Committee adopted the measures but implementation of each measure involves action by one or more groups. If the responsible groups implement all of the measures, the effect will be significant reductions in the impact of aircraft noise in the vicinity of Stapleton.

The general public has a continuing role in achieving the goals of the Plan. The public must keep officials aware of the need for continued effort to minimize adverse effects of aircraft noise.

In the nine months from completion of the Study until this paper was written, a number of steps had been taken to implement the Plan. The Mayor of Denver is establishing a committee to monitor noise abatement efforts. Denver has also requested the FAA to evaluate the opportunities for changing the departure routes and refining the preferential runway use system. The FAA is assessing the opportunities and the environmental impacts of these changes. Denver has been reviewing all of the land use options. This has included holding a public hearing on the problem. Other communities have been taking other actions. Denver is hiring a noise abatement officer for Stapleton.

During the presentation in May, I will give the current status of implementation of all elements in the Plan.

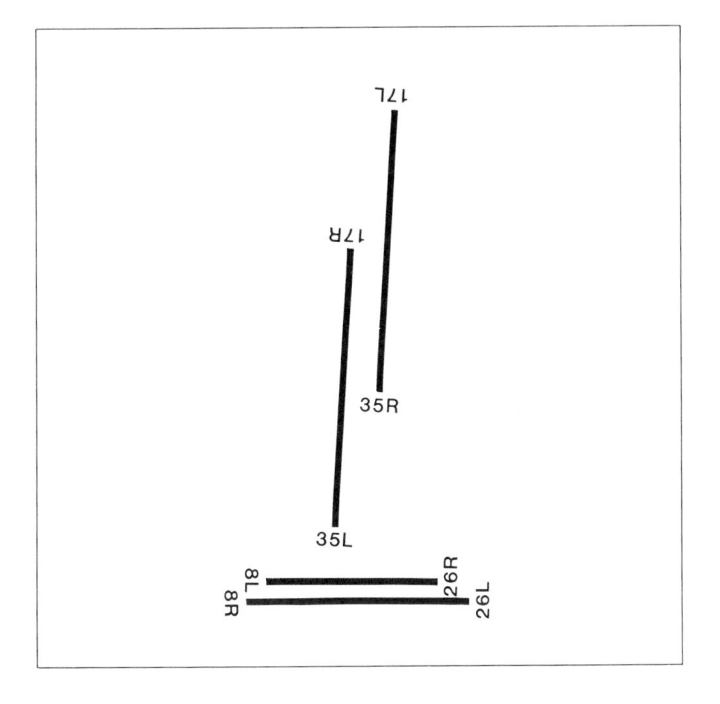

FIGURE 1. STAPLETON RUNWAY LAYOUT

REFERENCE

1. Harris Miller Miller & Hanson Inc.,"Final Report and Noise Mitigation Plan for Stapleton International Airport," Denver, CO, March 1984.

Noise Characteristics of Alternative Urban Transit Technologies

Carl E. Hanson*

This paper summarizes the noise characteristics of alternative transit technologies for an urban area. Among the transportation technologies considered are a busway system, light rail transit (LRT), rail rapid transit (RRT), intermediate capacity transit system (ICTS), and low-capacity transit system. Noise sources associated with operations of vehicles in each type of system are described and quantified for comparison among technologies.

INTRODUCTION

Metropolitan areas planning efforts toward solving traffic congestion problems include consideration of mass transit systems. A key element of the environmental impact assessment is the noise generated by the system. Experience has shown that noise is among the most cited of complaints against a new transit system. A transit system is, of necessity, routed where the people are living; it can be noisy, and it is often a new, unaccustomed source.

Trade-offs among alternative technologies include an early review of environmental effects. A major consideration in costing and routing of a new system is the need for inclusion of noise mitigation in adversely affected areas. A measure of the need for noise mitigation is provided by the distance to the threshold of impact for typical operations of the mass transit system.

NOISE FROM ALTERNATIVE TRANSIT VEHICLES

Bus Systems

Two types of buses are used in urban areas--a city bus and a commuter bus. Noise from each type depends on the operating characteristics. In general, engine noise dominates at low speed and tire/roadway noise dominates at cruising speed.

A city bus is used on local service routes in which low speeds and stop-and-go driving with frequent full-throttle acceleration is common. They typically have two axles and six tires. The dominant noise sources for city buses are:

- engine exhaust
- cooling fans
- engine casing.

*Vice President, Harris Miller Miller & Hanson Inc., 429 Marrett Road, Lexington, MA 02173

Cooling fans and radiators are located on the streetside of a bus, causing higher noise levels on that side. The exhaust noise is especially dominant during full acceleration away from a bus stop. Some makes of buses have high exhaust outlets which places the noise source at the top of the bus. This high exhaust placement increases the difficulty of using noise barriers as a mitigation measure. Bus noise is characterized by two tests: a "pull-away" test--simulating maximum noise levels during a pull away from the curb, and a constant speed pass-by test--simulating highway or street noise levels.

Typical noise levels from city buses are as follows[5, 3]:

- Pull-away noise level at 50 feet (15m):
  - --78 dBA on curbside
  - --83 dBA on streetside

- Pass-by noise level at 50 feet (15m):
  - --76 dBA at 30 mph
  - --79 dBA at 55 mph

A <u>commuter bus</u> is similar to an intercity long-haul bus. They usually have two axles but can have up to three axles and ten tires. Because a commuter bus is operated at high average speed on a highway or transitway, tire noise becomes more important than the sources on a city bus. Nevertheless, exhaust noise continues to dominate for acceleration at low speed.

The dominant noise sources for commuter buses are:

- tire/roadway interaction (highway speeds)
- engine exhaust
- cooling fans.

Like city buses, the streetside of a commuter bus has a higher noise level than the curbside. Typical noise levels from a commuter bus are[5]:

- Pull-away noise level at 50 feet (15m):
  - --82 dBA on curbside
  - --85 dBA on streetside

- Pass-by noise level at 50 feet (15m):
  - --77 dBA at 30 mph/48 kph
  - --80 dBA at 55 mph/88 kph.

<u>Rail Rapid Transit</u>

Pass-by noise characteristics of conventional steel-wheeled rail rapid transit are dominated by wheel/rail interaction at low and moderate speeds and propulsion systems at high speed (above 45 mph/ 72 kph).

The wheel/rail interaction includes the noise from small surface irregularities on the rail running surfaces (roar), impact noise from joints, turnouts and crossings (clickety-clack) and a high-pitched screeching noise from sharp curves (wheel squeal). In addition, the wheel/rail forces cause additional noise to be radiated from aerial guideway structures. Propulsion system noise includes cooling fan noise, gear noise, and electric motor casing. Noise from a rapid transit train increases with speed, with the dominant noise sources changing as speed increases. The dominant noise sources are as follows:

- Train stopping - brakes
- Train stopped - cooling fans on motors and air conditioners
- Train accelerating - traction motors
- Train at low speed (up to 45 mph/72 kph) - wheel/rail noise
- Train at high speed (above 45 mph/72 kph) - propulsion system.

Noise level also increases slightly with increase in train length but the chief effect of longer trains is the increase in noise exposure time due to the time it takes for a long train to pass.

The type of guideway structure can make a significant difference in the noise radiated from a train. Noise from trains at-grade is generally lower than from trains on aerial structure. Besides the fact that the aerial guideway places the noise source some 20 to 40 feet (6m to 12m) above the ground, the guideway structure itself radiates noise caused by the vibrations of the wheel/rail interaction. Track on ballast-and-tie is the most common type of guideway support for surface operations. Noise levels at 50 feet (15m) from an eight-car train at each of the two guideway configurations are as follows:

- Pass-by noise level at 50 feet (15m) from 8-car train at 50 mph/80 kph:
  - --82 dBA at-grade
  - --86 dBA aerial.

Light Rail Transit

Pass-by noise characteristics of light rail transit vehicles with steel wheels are similar to those of rapid transit cars under the same operating conditions. One difference is that LRT can be used in mixed traffic when a catenary-pantograph system is used for power collection. Under such conditions, the vehicles are operated at low speed and the noise levels are correspondingly low. Trains of up to six cars are operated on lines with heavy ridership. Under conditions of exclusive right-of-way, speeds and noise levels are comparable to rapid transit trains. The dominant noise sources for LRT are:

- wheel/rail interaction (especially squeal on tight curves)
- traction motors
- auxiliaries.

LRT is often placed in service in densely populated urban areas where tight turns are required for in-street operations. Vehicles with steel wheels on steel rails following curves with radii of less than 1,000 feet (300m) tend to squeal. Consequently, wheel squeal is one of the major noise problems of the LRT mode.

The type of guideway structure can make a significant difference in the noise radiated from LRT vehicles. Noise from trains at-grade tends to be lower than from trains on aerial structure. Besides the fact that the aerial guideway places the noise source well above the ground, the guideway structure itself radiates noise caused by the vibrations of the wheel/rail interaction. Noise from two-car LRT vehicles used in Calgary and Edmonton are as follows[2]:

- Pass-by noise at 50 feet (15m) from LRT trains at 50 mph/80 kph:
  - --two-car trains: 81 dBA
  - --four-car trains: 82 dBA

Low and Intermediate Capacity Transit

This transit mode encompasses a wide range of transportation vehicles providing local circulation in downtown areas, airports, and theme parks. Vehicle types are varied: monorail trains, rubber-tired vehicles, steel-wheeled intermediate capacity rail vehicles with linear induction motors, some with operators and some fully automatic. The wide variety of vehicle type makes it impossible to categorize the noise sources simply because each vehicle type is essentially unique. In general, pass-by noise levels from these vehicles are low due to low-speed operation. Those with rubber tires have the noise characteristics of automobiles where tire/roadway interaction is the dominant source at speed. A noticeable noise source for some of these vehicles is the control system. The electric motor-chopper speed controls emit a distinctive whine which can dominate the noise emission level from the vehicle at low speeds.

Monorail systems are distinguished by their preponderance of wheels: they have up to eight rubber tires for each suspension assembly with two suspensions per vehicle. This provides a great number of tire/guideway interaction noise sources. These tires are usually covered by skirts during operations on the guide rail resulting in adequate noise shielding.

In general, the dominant noise sources from low and intermediate capacity transit vehicles are:

- Propulsion systems, including controllers
- Ventilation systems
- Tire/guideway interaction.

Typical noise levels from a steel-wheeled ICTS two-car vehicle are as follows[6]:

- o Pass-by noise at 50 feet (15m) from ICTS vehicle (two-cars) at 50 mph/80kph:
  - --aerial guideway: 74 dBA (measured at guideway height)
  - --at-grade: 72 dBA (estimated from measurements at other speeds).

SUMMARY

A summary of noise characteristics from the various transit alternatives is shown in Table 1. The noise generated by typical operations of the vehicle for each system is compared in three ways: the pass-by noise level at a typical operating speed, the sideline distance to where the maximum noise level from a single vehicle reaches 75 dBA, and the sideline distance to where the equivalent noise level caused by the number of vehicles required to carry 300 passengers reaches 75 dBA. The 75 dBA level is used because it is the impact criterion used for average residential areas in the APTA Guidelines (Reference 1). The third measure is used in order to compare the vehicles on an equal basis. The results show that bus, RRT, and LRT modes are equivalent to each other in noise generation when compared with the same passenger loads. ICTS has a smaller noise impact area than the others.

TABLE 1. Comparison of Noise Effects of Alternative Technologies

					Distance to 75 dBA	
Technology	Number Passengers/ Vehicle	Guideway	Speed (mph)	Lmax * at 50 Feet	Single Vehicle	Vehicles with 300 Passengers
Bus	50	at-grade	50	78 dBA	80 ft.	270 ft.
Bus	50	aerial	50	80 dBA	90 ft.	370 ft.
RRT	150	at-grade	50	80 dBA	90 ft.	170 ft.
RRT	150	aerial	50	84 dBA	140 ft.	315 ft.
LRT	100	at-grade	50	79.5 dBA	85 ft.	215 ft.
LRT	100	aerial	50	83.5 dBA	135 ft.	400 ft.
ICTS	50	at-grade (NA)				
ICTS	50	aerial	50	72.5 dBA	40 ft.	116 ft.

* Maximum pass-by level for single vehicle even if it normally comes in a two-car married pair.

REFERENCES

1. American Public Transit Association, "1981 Guidelines for Design of Rail Transit Facilities", January 1979.

2. Andrew, C., D. L. Parsons, D. A. Towers, "A Comparison of Theoretical and Measured LRT Noise and Vibration: The Calgary Experience", Proceedings of Inter-Noise 1982, San Francisco, p. 209, May 1982.

3. Harris Miller Miller & Hanson Inc. measurements, April 1984.

4. Saurenman, H. J., J. T. Nelson, G. P. Wilson, Handbook of Urban Rail Noise and Vibration Control, U. S. Department of Transportation, UMTA-MA-06-0099-82-1, October 1982.

5. United States Environmental Protection Agency, Proposed Bus Noise Emission Regulation: Part 2. Background Document, EPA 550/9-77-201, August 1977.

6. Unsworth, G., Metro Canada Limited, private communication, May 1984.

Summary of the Accomplishments of the Federal Rail Transit Vibration and Noise Control Technology Program

Michael G. Dinning*

ABSTRACT

The U.S. Department of Transportation's Urban Mass Transportation Administration and Transportation Systems Center have played an active role in assisting the U.S. rail transit industry in developing, evaluating and promoting the implementation of improved vibration and noise control technology. This paper will review the major results of the federal program, and highlight examples of the implementation of these results by the transit industry.

Introduction

Improved technology to effectively control the environmental impact of rail transit noise and vibration *is* now available, and is being installed on both new and older rail transit systems. For the past decade, the U.S. Department of Transportation's Urban Mass Transportation Administration (UMTA) and Transportation Systems Center (TSC) have played an active role in assisting the U.S. transit industry in developing, evaluating and promoting the implementation of improved vibration and noise control technology. (4,5) In this presentation, I will review the major results of the UMTA/TSC Vibration and Noise Control Technology Program and highlight examples of the implementation of these results by the transit industry.** Technical details on the projects I will be describing may be found in the references given, available from TSC, or the National Technical Information Service.

Program Evolution

The Vibration and Noise Control Technology Program is one of the more mature R&D activities sponsored by UMTA. The program was established in the early 1970s during the formative years of the UMTA R&D programs, in response to a survey of transit operators in which noise control emerged as the most consistently cited topic in need of federally funded R&D.

*U.S. Department of Transportation, Research and Special Programs Administration, Transportation Systems Center, Kendall Square, Cambridge, MA 02142.

**Presentation accompanied by illustrative 35 mm slides.

Virtually every rail transit system in the U.S. has encountered significant noise and vibration problems, resulting in environmental, operational and economic impact. In New York City, for example, public concern led to the passage of the New York State Rapid Transit Noise Code, which requires the transit authority to develop and monitor noise abatement strategies.

Over the years, the federal program, working closely with the transit industry, has successfully directed activities through the entire sequence of technology development including:

1. Assessment of the problem;

2. Development of improved technology;

3. Evaluation of existing and prototype technology; and

4. Full scale implementation of improved technology.

The UMTA/TSC program has served as a focal point for noise and vibration control activities and information, nationwide. Particular emphasis has been placed on securing the involvement of the transit industry, including: transit authority engineers and planners, consultants, equipment suppliers and designers, professional societies, government agencies and even public interest groups. In many cases, cooperative studies have been conducted, with transit authorities and private sector investment complementing the federal R&D funds. Often, opportunities arose so that these projects were directed to help solve specific problems at a particular transit authority; the results were then integrated into the context of the overall program to advance the state-of-the-art of knowledge and technology.

The result has been a satisfying record of accomplishments in which program sponsored activities have led directly to the development and implementation of improved noise and vibration control technology on U.S. rail transit systems.

These activities have addressed the following major noise and vibration problems:

1. Wheel/Rail Noise;

2. Propulsion System Noise;

3. Brake System Noise;

4. Elevated Structure Noise;

5. Groundborne Noise & Vibration.

Wheel/Rail Noise Control

A comprehensive understanding of the generation of rail transit noise requires an understanding of the noise produced by wheel/rail interaction, for this is a major contributor to overall noise levels. (6,23) UMTA has sponsored a thorough program of research and development to develop an explanation of the mechanisms of wheel/rail noise generation, including:

1. Development of analytical models (9,10,15,16);

2. Laboratory and scale model tests;

3. Full-scale tests at the Transportation Test Center (TTC) in Pueblo, CO; and

4. In-service evaluative testing on U.S. transit systems. (2,3,7,11,19,20,24,25)

These studies produced an understanding of the relationship between noise levels and wheel/rail design and condition, resulting in recommendations such as: (17)

1. Criteria for minimum curve geometry to avoid squeal on new systems;

2. Guidelines for cost-effective application of wheel/rail maintenance techniques; and

3. Design parameters to "reinvent the wheel" for noise control purposes.

Currently, TSC is assisting the New York City Transit Authority (NYCTA) in evaluating the effectiveness of five types of transit car wheels fitted with damping treatments to eliminate the high-pitched squeal produced as trains negotiate tight curves. Damped wheels have been successfully adopted by the transit systems in Chicago (CTA) and New Jersey (PATCO). Based on the results of testing of damped wheels by TSC, the new Miami Metrorail vehicles include grooves for retrofit of damping rings should they be needed. TSC will distribute the final results of the NYCTA damped wheel tests widely, as all damping treatments appear to provide an effective means of eliminating one of the most irritating transit noise problems.

Propulsion System Noise Control

Tests conducted by TSC to evaluate various techniques to reduce wheel/rail noise revealed an unexpected problem; that noise from the vehicle propulsion system can dominate overall noise levels at typical operating speeds. Extensive diagnostic testing by TSC determined that significant noise reductions were possible by modifying the motor cooling fan and end housing. (1) Following these studies, suppliers of propulsion units in the United States then developed improved designs for their respective products. Under UMTA sponsorship, the NYCTA is now engaged in a program to evaluate the acoustical and cooling characteristics of the improved designs. Implementation on new vehicles and retrofits to the existing NYCTA vehicle fleet is planned.

Brake System Noise Control

Recently, the Washington Metropolitan Area Transit Authority (WMATA) encountered severe squeal and wear problems with their disc brakes. TSC assisted WMATA in assessing the magnitude and cause of the squeal problem. (18) Under a New Technology Introduction Program grant from UMTA, WMATA is working with the New York Air Brake Company to introduce a new, quiet, and less costly to maintain brake system design. The first production units were installed in May 1984; in total, 50 car sets will be monitored with regard to their acoustical and operational performance.

Elevated Structure Noise Control

Noise levels from older elevated track structures are typically over 90 dB, representing perhaps the most serious transit noise impact. (26) Testing by TSC has determined that newer, higher mass structures utilizing continuous welded rail are considerably quieter, and can be improved further by use of barrier walls and structural damping. (27) TSC has directed substantial research on the very difficult problem of the abatement of noise from older steel structures. Complex analytical modelling and field tests were conducted to determine how noise is generated from these structures. Design guidelines for noise control from steel structures were produced. (28) A subsequent effort, conducted in cooperation with the rail fastener manufacturing industry, examined the feasibility of noise reduction by a modification of resilient track fastener design. (8) The NYCTA has received an UMTA grant to apply the results of these analytical studies and evaluate various types of resilient rail fasteners on the elevated track structure. The fasteners will be evaluated with regard to their noise abatement and operational performance. The objective of this demonstration is to identify a fastener which could be installed as part of structure rehabilitation programs, now underway in New York and Chicago.

Prediction and Control of Groundborne Vibration

The environmental impact of groundborne noise and vibration from transit operations is another extremely complex area in which the UMTA/TSC program has sponsored research to advance the state-of-the-art in mitigation technology. Design guidelines for floating slab trackbed produced by the Program in 1975 aided many new transit systems in eliminating the groundborne vibration problem. (12) However, continued occurrences of community complaints and lawsuits regarding vibration, and the high cost of mitigation, suggested that a better understanding of the phenomenon was needed. After an extensive review of international research and experiences, TSC initiated development of an improved vibration prediction methodology. (13,14) Because of the difficulty in modelling the propagation of vibration, this approach is based on field tests of propagation efficiency in the proposed right-of-way, using a calibrated impact source. The procedure has been exercised in trial applications for the planned rail systems in Los Angeles, Baltimore and San Jose with very promising results. By more accurately specifying the extent of vibration control technology needed, significant savings in system construction costs are possible.

To complement the field measurements of vibration propagation, TSC is compiling a catalog of vibration source spectra for various combinations of vehicles and track support systems, both of which will have an effect on the nature of the vibrations produced. Data has been gathered during tests at the DOT Transportation Test Center in Pueblo, CO for the vehicles used on the Atlanta, Buffalo, Portland, Miami and Baltimore transit systems. In addition, vehicles have been evaluated in Boston, San Diego, and San Francisco. These vehicles represent a wide range of suspension stiffnesses, a factor which has been shown to influence significantly the ground vibration produced.

A Vibration Prediction Users Manual will be prepared outlining the new prediction procedures and presenting data showing the influence of vehicle and track system design, soil type, and propagation distance on vibration levels.

UMTA and TSC are also involved in efforts to evaluate vibration control techniques. The Cologne Egg, a resilient rail fastener designed primarily for vibration control, has been evaluated for one year in revenue service operations on the Massachusetts Bay Transportation Authority in Boston, under UMTA sponsorship. The effect of modifications to vehicle suspension systems has also been evaluated in cooperation with the Metropolitan Atlanta Rapid Transit Authority. In addition, ground vibration impacts will be a consideration in the development of transit truck design guidelines now underway at TSC.

Information Exchange

A major role of the UMTA/TSC Vibration and Noise Control Technology Program has been to facilitate exchange of information between the many organizations involved in developing or using noise and vibration control technology. Indeed, interaction among all those involved in this process is essential to the success of the program in reducing noise and vibration impact most cost effectively. Various communications mechanisms have been used, including: conferences, workshops, press releases to trade journals and technical publications, direct mailing of DOT reports, American Public Transit Association (APTA) Liaison Board meetings, and technical briefings to transit managers, engineers and planners throughout the country. Briefings often featured an audio-visual presentation, _Urban Rail Noise Control_, which gives an overview of how to address noise and vibration problems as part of transit system design and maintenance programs.

The UMTA/TSC program has supplied technical information to every existing rail transit system in the country and to many of those still in the planning stages. The most comprehensive reference on the subject is the _Handbook of Urban Rail Noise and Vibration Control_. (21,22) Since we published the Handbook in 1982, it has been used by transit system engineers and designers throughout the world, from Santa Clara to Singapore, to apply noise and vibration control technology in the design and rehabilitation of rail transit systems.

Future Activities

The Program is currently completing ongoing research and analysis activities so that the results can be disseminated to potential users or integrated into broader technical research topics. The results of our activities to develop improved ground vibration prediction procedures, and to evaluate damped wheels, will be documented, distributed, and presented at DOT-sponsored workshops. The information generated on the control of elevated structure noise, propulsion system noise and ground vibration will be integrated into ongoing efforts sponsored by UMTA to develop or evaluate improved technology for structure rehabilitation and vehicle design.

The results of the UMTA/TSC Vibration and Noise Control Technology Program are now being implemented as part of over one billion dollars worth of capital and rehabilitation programs on U.S. transit systems throughout the country. By continued integration of research results into operational design, rehabilitation and maintenance programs, potential exists to realize additional environmental benefits for the transit rider and wayside community, and operating and economic cost savings for the transit system operator.

REFERENCES

1. Dixon, N.R., L.G. Kurzweil, and P.J. Remington, "Propulsion Systems Noise in Urban Rail Transit Systems," April 1983. (Limited distribution)

2. Holowaty, M.C., H.J. Saurenman, and S.M. Rosen, "In-Service Performance and Costs of Methods for Control of Urban Rail System Noise - Experimental Design," May 1976. UMTA-MA-06-0025-76-4, PB-257-200.

3. Holowaty, M.C., and H.J. Saurenman, "In-Service Performance and Costs of Methods to Control Urban Rail System Noise - Test and Evaluation Plan," April 1977. UMTA-06-0025-77-10, PB-272-521.

4. Kurzweil, L.G., W.N. Cobb, and M.G. Dinning, "Urban Rail Noise Abatement Program A Description," U.S. DOT, Transportation Systems Center, March 1980. UMTA-MA-06-0099-80-2.

5. "Urban Rail Noise Abatement Program Digest," Transportation Systems Center, July 1981. UMTA-MA-06-0099-80-3.

6. Kurzweil, L.G., and L.E. Wittig, "Wheel/Rail Noise Control - A Critical Evaluation (Interim Report)," January 1981. UMTA-MA-06-0099-81-1.

7. Kurzweil, L.G., "In-Service Testing of Wheel/Rail Noise Control Treatments," paper presented at the 1978 APTA Rapid Transit Conference, Chicago, June 1978, U.S. DOT, Transportation Systems Center.

8. Kurzweil, L.G., "Design of a Resilient Rail Fastener for Minimizing Noise from an Open Wood-Tie Deck, Steel Plate Stringer Elevated Structure," November 1982. (Limited distribution)

9. Lotz, R., "Wheel/Rail Noise: A Progress Report" in "INTER-NOISE 74," Proceedings of the 1974 International Conference on Noise Control Engineering, U.S. DOT, Transportation Systems Center, October 1974.

10. Lotz, R., "Wheel/Rail Noise Reduction in Practice" in "INTER-NOISE 77," Proceedings of the International Conference on Noise Control Engineering, U.S. DOT, Transportation Systems Center, March 1977.

11. Lotz, R., "Railroad and Rail Transit Noise Sources" in the "Journal of Sound and Vibration Control," Vol. 51, No. 3., U.S. DOT, Transportation Systems Center, April 8, 1977.

12. Manning, J.E., D.C. Hyland, and G. Tocci, "Vibration Prediction Models for Floating Slab Rail Transit Track," August 1975. UMTA-MA-06-0025-75-13, PB-245-638.

13. Nelson, J.T., H.J. Saurenman and T.A. Mugglestone, "State-of-the-Art Review: Prediction and Control of Groundborne Noise and Vibration from Rail Transit Trains--Annotated Bibliography," May 1982. UMTA-MA-06-0099-82-3, PB-83-100-420.

14. Nelson, J.T., and H.J. Saurenman, "State-of-the-Art Review: Prediction and Control of Groundborne Noise and Vibration from Rail Transit Trains," December 1983. UMTA-MA-06-0049-83-4.

15. Remington, P.J., M.J. Rudd, and I.L. Ver, "Wheel/Rail Noise and Vibration Control (Final Report), Vol. I," May 1975. UMTA-MA-06-0025-75-10, PB-244-514.

16. Remington, P.J., M.J. Rudd, and I.L. Ver, "Wheel/Rail Noise and Vibration Control (Final Report), Vol II," May 1975. UMTA-MA-06-0025-75-11, PB-244-515.

17. Remington, P.J., N.R. Dixon, and L.E. Wittig, et. al., "Control of Wheel/Rail Noise and Vibration," April 1983. UMTA-MA-06-0099-82-5.

18. Ronk, L.A., and M.A. Staiano, "Evaluation of Squeal Noise from the WMATA Transit Car Disc Brake System: A Preliminary Investigation (Final Report)," March 1981. UMTA-MA-06-0099-81-4.

19. Saurenman, H.J., "In-Service Performance and Costs of Methods to Control Urban Rail System Noise - Second Test Series Report," October 1979. UMTA-MA-06-0099-79-4.

20. Saurenman, H.J., G.P. Wilson, and R.L. Shipley, "In-Service Performance and Costs of Methods to Control Urban Rail System Noise Final Report," December 1979. UMTA-MA-06-0099-80-1.

21. Saurenman, H.J., J.T. Nelson, and G.P. Wilson, "Handbook of Urban Rail Noise and Vibration Control," February 1982. UMTA-MA-06-0099-82-1, PB82-220-757.

22. Saurenman, H.J., J.T. Nelson, and G.P. Wilson, "Handbook of Urban Rail Noise and Vibration Control: Executive Digest," February 1982. UMTA-MA-06-0099-82-2, PB82-220-427.

23. Schultz, T.J., "Development of an Acoustic Rating Scale for Assessing Annoyance Caused by Wheel/Rail Noise in Urban Mass Transit," February 1974. UMTA-MA-06-0025-74-2, PB-231-363.

24. Shipley, R.J., and H.J. Saurenman, "In-Service Performance and Costs of Methods to Control Urban Rail System Noise - Initial Test Series Report," August 1978. UMTA-MA-06-0025-78-7, PB-288-838.

25. Slutsky, S., W.R. McShane, and J.J. Starace, "Noise Degration Over Time in Rail Rapid Transit Cars," December 1978. UMTA-NY-11-0002-79-1, PB-292-031.

26. Towers, D.A., "Noise Impact Inventory of Elevated Structures in U.S. Urban Rail Rapid Transit Systems (Interim Report)," September 1980. UMTA-MA-06-0099-80-5, PB81-120-958.

27. Ungar, E., and E. Wittig, "Wayside Noise of Elevated Rail Transit Structures: Analysis of Published Data and Supplementary Measurements (Interim Report)," December 1980. UMTA-MA-06-0099-80-6.

28. Ungar, E.E., "Design Considerations for Retrofit Quieting of Elevated Rail Transit Structures with Open Wood-Tie Decks on Steel Stringers," January 1982. (Limited distribution)

Noise Assessment, Measurement, and Mitigation, an OSHA View

Raymond G. Kunicki

A discussion is presented on the Occupational Safety and Health Administration (OSHA) 29 CFR 1910.95 General Industry noise exposure standards and the 29 CFR 1926.52 and 1926.101 Construction Industry noise exposure and hearing protection standards. Clarification is made on the OSHA mandate to assess and measure employee exposure to noise versus environmental or area noise due to specific sources.

This is the third session of this symposium, a session on stationary source noise and vibration measurement, prediction, and mitigation. With your indulgence, I'd like to diverge a bit with my presentation and talk about the assessment, measurement, and mitigation of noise from the Occupational Safety and Health Administration (OSHA) point of view - one that encompasses construction noise, transportation noise, stationary noise, mobile noise, in fact all possible man-made noise in a workplace affecting employees in that workplace.

A glance at the titles of papers presented at these sessions tells one that the majority of papers are concerned with noise sources, construction, transportation, or stationary, affecting an environment. Usually a concern is raised about annoyance of the general population or annoyance of neighbors. Sometimes an immediate environment is of concern - witness the articles concerning the effects of vibration on plant processes. Many times the annoyance is severe and can run afoul of local property line or emissions ordinances or state statutes or federal regulations.

The OSHA point of view (or point of listening, if you prefer) of the assessment, measurement, and mitigation of noise sources, is interested exclusively in a very local environment: the noise levels in the vicinity of an employee's ear. OSHA is also concerned with very high noise levels that go beyond annoyance, indeed, levels that can cause permanent hearing loss.

U.S. Public Law 91-596, enacted on December 29, 1970 -

Author's Address:
9103 Grant Avenue
Manassas, Virginia 22110

the Occupational Safety and Health Act of 1970 - orders OSHA "...to assure safe and healthful working conditions for working men and women". To this end, OSHA adopted the 29 CFR (Code of Federal Regulations) 1910.95 General Industry standards for occupational noise exposure from the Walsh-Healey Public Contracts Act standards of the U.S. Department of Labor Workplace Standards Administration. The Walsh-Healey standards were adopted in turn from the American Conference of Governmental Industrial Hygienists Threshold Limit Values for Physical Agents. (Copies of the OSHA 29 CFR 1910.95 General Industry noise exposure standards and the 29 CFR 1926.52 and 1926.101 Construction Industry noise exposure and hearing protection standards are appended to this paper. The standards, while not identical, contain the same basic provisions and criteria.)

Table G-16 of 29 CFR 1910.95 or Table D-2 of 29 CFR 1926.52 lists the maximum permissible noise exposures as functions of continuous sound levels and durations per day in hours. Loss of hearing occurs as a result of the cumulative effect of exposure to sound above a maximum intensity and over a maximum duration in a given period of time. OSHA's criterion has been set at a maximum level of 90 dB(A) for 8 hours out of a day. The amount of sound energy absorbed during such an exposure is considered to be the upper limit of a daily dose which will not produce disabling loss of hearing in more than 20% of the exposed population. Note that exposure to a sound level that is increased by five decibels halves the allowable exposure time. Thus, exposure to a continuous level of 95 dB(A) is permitted for four hours out of a day. A "dose" of 100% corresponds to an exposure at 90dB(A) for 8 hours. Similarly, 400% would be equivalent to an exposure at 100 dB(A) for 8 hours, and 50%, an 8-hour exposure at 85 dB(A). This is the same as expressing "F_6" of 1926.52 (d)(2)(ii) in terms of percentages where 1 = 100%.

OSHA Assessment

OSHA assesses noise levels by measurements at an employee's ear and demands by law mitigation for those same levels if they exceed the prescribed standard limits, i.e., the "dose" is greater than 100%. Note there is a lack of reference to any assessment or measurement of any noise source anywhere in the appended standards. This is important. OSHA has no authority to regulate noise <u>emissions</u> from noise sources. OSHA's mandate is to regulate <u>overexposures</u> to employees in workplaces. An OSHA citation will refer to an overexposed individual's job title but not to any source that contributed to the cited overexposure. It is indeed possible for an employee in a given workplace environment to be exposed to a variety of noise sources (construction, transportation, mobile, stationary, etc.) arriving through a variety of paths as part of a typical daily noise exposure. Moreover, it is possible for this to occur even with a relatively sedentary employee.

Measurement

The very nature of the OSHA standards specifying permissible continuous levels for varying amounts of time dictates that some sort of "averaged" exposure must be determined. In fact, this is the aforementioned "dose". This can be done manually with a sound level meter, stopwatch, notebook, and lots of patience, or it can be done automatically with an integrating sound level meter or noise dosimeter.

The OSHA standards also specify that these permissible continuous levels averaged over time are permissible personal exposures. This means that the above "dose" must be determined at an employee's local hearing zone for the entire exposure period. In the parlance of industrial hygiene this is called personal monitoring. Monitoring of local workplace areas without regard to employee presence or movement is called area monitoring. Area monitoring is usually less time consuming than personal monitoring and affords a better description of a workplace environment for noise source and path assessment purposes.

OSHA uses noise dosimeters and personal monitoring in its enforcement procedures. However, a workplace with well-defined continuous noise levels and relatively immobile employees could easily nominate itself as a good candidate for any of the various area monitoring techniques. In this case, the doses calculated from the area monitoring results should agree quite well with doses recorded through personal monitoring.

Mitigation

The most frequently cited (and contested) OSHA health citation over the past decade is 29 CFR 1910.95 (b)(1) - failure to implement all feasible engineering and administrative controls for excessive noise levels. Administrative controls are defined as those controls that reduce an employee's exposure time - usually through employee rotation. Engineering controls are those that affect the noise sources and paths. So, after OSHA assesses noise levels by measurements and demands mitigation for excessive levels noted at an employee's ear, it is necessary to continue with a further assessment of the sources and paths contributing to these excessive levels (more measurements) and make a determination what the mitigation should be to reduce these excessive levels. The onus of this determination lies with the employer, although OSHA usually does offer suggestions for mitigation. (In a non-enforcement vein, free on-site technical advice may be obtained from OSHA-supported state and private consultation organizations. Priority is given to the needs of small business. Contact the OSHA Office of Consultation, Room N3472, 200 Constitution Ave., N.W., Washington, DC 20210. Phone: (202)5237266.) Clearly, this further assessment is accomplished using noise source and path analysis rather than OSHA enforcement dose measurement. This is an assessment/measurement situation that would be more familiar to a design, construction, or environmental engineer.

Final mitigation of excessive noise exposure can be accomplished with administrative and/or engineering controls. Administrative controls are usually impractical due to employee mobility or restrictions applied to job duties and classes through bargaining agreements. Engineering controls implemented as a result of OSHA citations are intrinsically retrofit controls to sources and paths. (Recall that OSHA has no authority to regulate equipment noise emissions.) This usually leads to the incorporation of the two basic tenets of noise control engineering:

1. Put a box on it (the source).
2. Spray the ceiling.

If noise control were considered prior to facility construction, a good deal of this kludge-like design work (and its attendant waste of money) would be avoided. This pre-construction consideration would include both facility design and the use of quieter machinery and equipment.

Curiously, in 1971 OSHA issued its Bulletin 334, _Guidelines to the Department of Labor's Occupational Noise Standards_, that discussed noise control measures:

"The following are examples [of engineering controls]:

A. Maintenance
   1. Replacement or adjustment of worn and loose or unbalanced parts of machines.
   2. Lubrication of machine parts and use of cutting oils.
   3. Properly shaped and sharpened cutting tools.

B. Substitution of Machines
   1. Larger, slower machines for smaller, faster ones.
   2. Step dies for single operation dies.
   3. Presses for hammers.
   4. Rotating shears for square shears.
   5. Hydraulic for mechanical presses.
   6. Belt drives for gears.

C. Substitution of Processes
   1. Compression for impact riveting.
   2. Welding for riveting.
   3. Hot for cold working.
   4. Pressing for rolling or forging.

D. Vibration Dampening
   1. Increase mass.
   2. Increase stiffness.
   3. Use rubber or plastic bumpers or cushions.
   4. Change size to change resonance frequency.
E. Reducing Sound Transmission Through Solids
   1. Flexible mountings.
   2. Flexible sections in pipe runs.
   3. Flexible shaft couplings.
   4. Fabric sections in ducts.
   5. Resilient flooring.
F. Reducing Sound Produced by Fluid Flow
   1. Intake and exhaust mufflers.
   2. Fan blades designed to reduce turbulence.
   3. Large, low speed fans for smaller, high speed fans.
G. Include Noise Level Specifications When Ordering New Equipment
H. Isolating Noise Sources
   1. Completely enclose individual machines.
   2. Use baffles.
   3. Confine high noise machines to insulated room.
I. Isolating Operator
   Provide a relatively sound-proof booth for the operator or attendant of one or more machines."

Now, nearly 15 years later, most of the controls that are employed are still of the remedial variety.

OSHA insists that engineering controls shall be applied even if the exposure limits of Table G-16 of 29 CFR 1910.95 are still exceeded after application of said controls provided a significant reduction can be shown in an employee's exposure. A significant reduction is defined as a reduction of three or more decibels. Note that this is an exposure reduction and not necessarily a reduction of source emissions. An employee enclosure or booth is considered to be an engineering control since it removes the employee from noise paths. (However, an employee must spend one third of his work shift in a virtually exposureless enclosure to obtain a reduction of three decibels in his overall exposure!) Ear muffs and inserts are considered to be personal protective equipment and must be used according to 29 CFR 1910.95 (b)(1), to reduce exposures to within the levels of Table G-16 if engineering and administrative controls fail to do so.

Summation

I'd like to leave you with some points to remember concerning OSHA and occupational noise exposure:

1. OSHA is only concerned with employee exposure to high workplace noise levels, levels that can cause permanent hearing loss.
2. OSHA uses noise dosimetry and personal sampling in its assessment and enforcement of employee noise exposure.
3. OSHA cannot require machinery and equipment manufacturers to meet any noise emission requirements for new machinery and equipment. As a result, OSHA-mandated engineering controls are usually after-the-fact makeshift controls.
4. Therefore, it is hoped that employee noise exposure is one of the primary decision criteria that will be employed (by members of this audience, for example) in the design of the American workplace.

1910.95—OCCUPATIONAL NOISE EXPOSURE

(a) Protection against the effects of noise exposure shall be provided when the sound levels exceed those shown in Table G-16 when measured on the A scale of a standard sound level meter at slow response. When noise levels are determined by octave band analysis, the equivalent A-weighted sound level may be determined as folows:

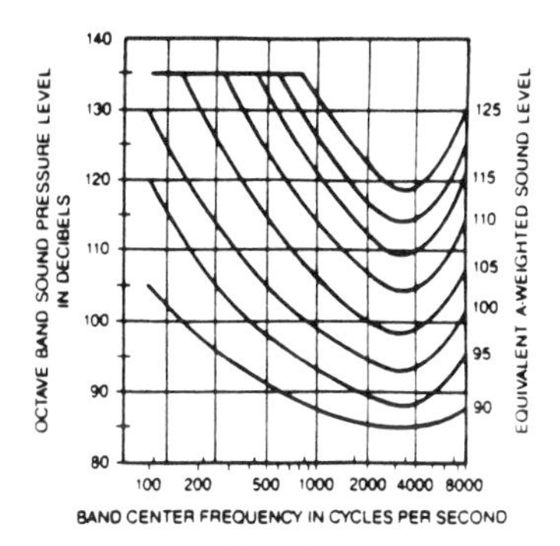

Figure G-9

Equivalent sound level contours. Octave band sound pressure levels may be converted to the equivalent A-weighted sound level by plotting them on this graph and noting the A-weighted sound level corresponding to the point of highest penetration into the sound level contours. This equivalent A-weighted sound level, which may differ from the actual A-weighted sound level of the noise, is used to determine exposure limits from the Table G-16.

1910.95—

(b)(1) When employees are subjected to sound exceeding those listed in Table G-16, feasible administrative or engineering controls shall be utilized. If such controls fail to reduce sound levels within the levels of Table G-16, personal protective equipment shall be provided and used to reduce sound levels of the table.

(2) If the variations in noise level involve maxima at intervals of 1 second or less, it is to be considered continuous.

(3) In all cases where the sound levels exceed the values shown herein, a continuing, effective hearing conservation program shall be administered.

Table G-16—Permissible Noise Exposures[1]

Duration per day, hours	*Sound level dBA slow response*
8	90
6	92
4	95
3	97
2	100
1½	102
1	105
½	110
¼ or less	115

[1]When the daily noise exposure is composed of two or more periods of noise exposure of different levels, their combined effect should be considered, rather than the individual effect or each. If the sum of the following fractions: $C_1/T_1 + C_2/T_2 C_n/T_n$ exceeds unity, then, the mixed exposure should be considered to exceed the limit value. *Cn* indicates the total time of exposure at a specified noise level, and *Th* indicates the total time of exposure permitted at that level.

Exposure to impulsive or impact noise should not exceed 140 dB peak sound pressure level.

1926.52—OCCUPATIONAL NOISE EXPOSURE

(a) Protection against the effects of noise exposure shall be provided when the sound levels exceed those shown in Table D-2 of this section when measured on the A-scale of a standard sound level meter at slow response.

(b) When employees are subjected to sound levels exceeding those listed in Table D-2 of this section, feasible administrative or engineering controls shall be utilized. If such controls fail to reduce sound levels within the levels of the table, personal protective equipment as required in Subpart E, shall be provided and used to reduce sound levels within the levels of the table.

(c) If the variations in noise level involve maxima at intervals of 1 second or less, it is to be considered continuous.

(d)

(1) In all cases where the sound levels exceed the values shown herein, a continuing, effective hearing conservation program shall be administered.

TABLE D-2—PERMISSIBLE NOISE EXPOSURES

Duration per day, hours:	*Sound level dBA slow response*
8	90
6	92
4	95
3	97
2	100
1½	102
1	105
½	110
¼ or less	115

(2)

(i) When the daily noise exposure is composed of two or more periods of noise exposure of different levels, their combined effect should be considered, rather than the individual effect of each. Exposure to different levels for various periods of time shall be computed according to the formula set forth in subdivision (ii) of this subparagraph.

(ii)

$$F_6 = \frac{T_1}{L_1} + \frac{T_2}{L_2} + \cdots + \frac{T_n}{L_n}$$

where:

F_6 = The equivalent noise exposure factor.
T = The period of noise exposure at any essentially constant level.
L = The duration of the permissible noise exposure at the constant level (from Table D-2).

If the value of F_6 exceeds unity (1) the exposure exceeds permissible levels.

(iii) A sample computation showing an application of the formula in subdivision (ii) of this paragraph is as follows. An employee is exposed at these levels for these periods:

110 dba ¼ hour.
100 dba ½ hour.
90 dba 1½ hour.

$$F_6 = \frac{1/4}{1/2} + \frac{1/2}{2} + \frac{1\,1/2}{8}$$

$F_6 = 0.500 + 0.25 + 0.188$
$F_6 = 0.938$

Since the value of F_6 does not exceed unity, the exposure is within permissible limits.

(e) Exposure to impulsive or impact noise should not exceed 140 dB peak sound pressure level.

1926.101 — HEARING PROTECTION

(a) Wherever it is not feasible to reduce the noise levels or duration of exposures to those specified in Table D–2, Permissible Noise Exposures, in §1926.52, ear protective devices shall be provided and used.

(b) Ear protective devices inserted in the ear shall be fitted or determined individually by competent persons.

(c) Plain cotton is not an acceptable protective device.

Electric Utility Noise Emissions

Allan M. Teplitzky*

Environmental regulations require that impact statements for new electric utility and industrial power plants include an evaluation of noise emissions. Variations in equipment sound power levels and the non-uniformity of state and local noise abatement regulations require that sound emissions from each proposed utility facility be uniquely evaluated. Acceptable community noise environments are achieved when: (a) emissions from new facilities are compatible with surrounding land use, (b) sufficient real estate is acquired for a buffer zone, and (c) noise abatement devices are installed. Thus, an understanding of both the regulatory requirements and the ability to identify practical and economical solutions for abating power plant noise emissions is needed by design engineers and environmental specialists.

DESIGN CRITERIA

Absent uniform national noise regulations, many state and local governments have promulgated permissible environmental noise limits. These regulations generally specify permissible sound levels measured either at a property-line or at the nearest neighbor. These typically range between 45 and 65 dB(A). Variations are attributable to local factors, such as the level of annoyance considered acceptable, zoning, land use, ambient noise levels, community needs, and compliance costs. When local or state noise limits have not been promulgated, guidance for acceptable community noise emissions can be obtained from the literature [1, 2, 3, 4, 5, and 6] and from recommendations published by government agencies [7].

Anticipated community reaction to the noise emitted from a power plant cannot be determined solely on the basis of comparing the expected sound levels with permissible limits [2]. Other factors which can influence community annoyance include the temporal and spectral frequency characteristics of the noise emissions, and the ambient sound level prior to construction. Experience indicates that electric utility facilities emitting discrete tones,

*Manager, Acoustics, Consolidated Edison Company of New York, Inc., 4 Irving Place, New York, NY 10003.

such as transformers, or impulsive type noises, such as steam venting, will have a greater environmental impact than power plants emitting noise with a continuous broad-band frequency spectra.

COMMUNITY NOISE EXPOSURE

Community exposure to power plant noise can be reduced either by installing noise abatement devices which abate the noise at the source or by acquiring sufficient real estate for an "acoustical buffer zone" between the plant and the surrounding community [8]. When real estate costs are high or an adequate buffer zone is not available, engineering solutions are the only feasible alternative. Equipment layouts that take advantage of the sound attenuation provided by local topography and flora, shielding provided by structures on the site and equipment orientation that takes advantage of sound-source directivity will minimize noise mitigation costs. Sources such as natural draft cooling towers or coal handling equipment that do not readily lend themselves to engineered noise abatement treatments require adequate buffer zones to minimize noise impacts [9, 10, 11].

Principle power plant environmental noise sources are (rank-ordered by A-weighted sound power level): forced draft fan inlets, steam safety valves and vents, induced draft fan discharges, gas recirculating fan casings, coal car shakers, boiler feed pumps and cooling towers [12]. Other equipment that may be judged annoying by the community because of special temporal or tonal characteristics include: transformers, public address systems, coal handling equipment and precipitator rappers [12]. From this list, equipment most likely to become significant community noise problems can be identified and appropriate noise abatement priorities are then established (Table I).

A procedure for predicting noise emissions from a power plant to the surrounding community has been developed [13]. Equipment operating characteristics and arrangement, and the influence of local environmental, topographical and meteorological factors on sound propagation are included in the procedure. Application of this procedure results in a contour map of the yearly average day-night A-weighted sound levels (Ldn) around a proposed plant. However, sound levels in the community as a result of power plant operations vary continuously due to changes in operating conditions and daily changes in meteorology, atmospheric sound absorption, and seasonal changes in acoustical buffer zone vegetation.

Table I

Typical Sources of Environmental Noise

Source	Temporal Pattern	A-Weighted Emission Level, dB (A) Lw=Sound Power Lp=Sound Pressure
Transformers	Continuous	Lp=55+12log(MVA) (No Specified Noise Control)
Cooling Tower		
- Mech. Draft	Continuous	Lw=85+10log(fan HP)
- Nat. Draft	Continuous	Lw=117 at rim for 400,000 GPM flow
Induced Draft Fan or Forced Draft Fan	Continuous	Lw=37+10log(Q) + 20log(SP) (Subtract 5 dB for FD fans with inlet control vanes)
Steam Vents	Intermittent	Lw=164 for a steam flow of 550 m/sec from a 3 in. pipe
Coal Handling Equipment	Intermittent	Lw=116 to 132 depending on equipment
Precipitator Rapper and Vibrators	Periodic	Lw (Rapper) = 111 Lw (Vibrator) = 114
Outdoor PA System	Intermittent	Lw=108 (average) Lw=120 (peak)

NOISE CONTROLS

Noise abatement requirements for each power plant are unique because of variations in electric power output, equipment type, plant layout, operating procedures and site conditions. Although there are general techniques that apply to all power plants (Table II), noise abatement modifications must be engineered to reflect these local conditions.

For existing plants, noise abatement is generally achieved by add-on devices, such as enclosures, mufflers or lagging. These retrofits are engineered to reflect job-site conditions and noise attenuation requirements. Enclosures are often selected for equipment such as boiler feed pumps, because they significantly reduce equipment noise and are easily installed. However, they may require special lighting and ventilation, are subject to damage, do not always protect employees working inside them, and may be resented by employees when they interfere with equipment maintenance. For large air moving devices, such as induced draft fans, mufflers that attenuate noise without excessive pressure loss are installed. These devices are designed to maximize sound attenuation at the blade passage frequency of the fan and to survive the harsh flue gas environment [14] .

For new plants, noise abatement is best achieved by specifying and procuring equipment with noise emission levels that insure the in-plant and environmental sound levels meet current or anticipated standards. Simply specifying that noise emissions should not exceed a specified standard is generally insufficient because of ambiguities in legal interpretations. In some cases, vendors will be unable to supply equipment, such as forced-draft fans, with noise emissions below the specified limit, unless add-on noise control devices, such as inlet mufflers, are installed. Even for new plants, add-on noise control devices will be needed for some equipment.

COST OF NOISE CONTROL

While noise abatement does not have the same financial impact on a power plant as air or water pollution controls, it does affect the selection of almost every piece of equipment, the design of piping and ducting systems, and the overall plant layout. Although noise abatement costs depend on local regulations, proximity of neighbors, plant layout, boiler fuel and condensor cooling system, these costs can be minimized by judicious equipment siting.

Table II

Achievable Noise Reductions From Add-On Modifications

Noise Control	Noise Reduction dB(A)					
	5	10	15	20	25	30
Barrier (outdoors)	Yes	Yes	Yes	?	No	No
Partial Enclosure	Yes	Yes	?	No	No	No
Enclosure	Yes	Yes	Yes	Yes	Yes	Yes
Sound Absorption (Large Room)	Yes	Yes	?	No	No	No
Lagging	Yes	Yes	Yes	?	?	?
Mufflers	Yes	Yes	Yes	Yes	Yes	Yes
Vibration Damping	Yes	Yes	?	?	No	No

Note: Yes - Achievable
No - Not achievable
? - Requires special design or can be achieved only in limited frequency range.

The following examples provide order-of-magnitude noise control cost estimates:

- A large architect/engineering firm estimated in 1976 that the cost to comply with the OSHA occupational noise standard and environmental noise regulations for a new power plant could range from 0.8% to 1.4% of the total plant cost for a fossil-fired plant, and from 0.1% to 0.25% for a nuclear plant [15].

- One European utility reported that it paid 4.5% of the total plant cost to reduce environmental noise to 52 dB(A) at 150 m. from three 150-MW units. Equipment at the plant included mechanically assisted natural-draft cooling towers [9].

CONSTRUCTION NOISE

Environmental impact assessments for new power plants generally include a construction noise analysis. Field measurements made at several power plant construction sites provide a basis for this analysis [16].

Noise emissions from power plant construction activities may be represented by a set of equi-sound level contours that are essentially circular in shape with the acoustic center generally located at the turbine generator. The equivalent A-weighted sound levels (Leq(A)) of each noise contour for each of the six major construction phases are listed in Table III. Diesel engines powering mobile construction equipment are the primary noise source. The estimated accuracy of the sound contours are about 3 to 5 dB, and the contours are based on the following assumptions: 59°F temperature, 70% relative humidity, no wind temperature gradients, flat terrain, and no vegetation losses.

The contours represent the estimated equivalent sound level for an entire construction phase. Short-term levels will obviously be either higher or lower than the average, especially during periods of atypical activities at the site.

Although the typical noise contours provide planning guidance, site specific noise emission contours that more accurately reflect the actual equipment being used, topography, meteorology, construction schedule, and equipment usage factors can be developed. However, this specific information is not always available when the environmental impact analysis is prepared. Therefore, power plant construction noise contours, which are estimated by the described method, provide a reasonable prediction of construction noise levels from a proposed

TABLE III

Equivalent Sound Levels (dBA) for Each Construction Phase

Construction Phase	Distance from the turbine-generator to the sound level contour, feet				
	2,400	3,400	4,800	6,700	9,100
Excavation	55	50	45	40	35
Concrete Pouring	51	46	41	36	31
Steel Erection	55	50	45	40	35
Mechanical	50	45	40	35	30
Clean-up	45	40	35	30	25
Steam Blow-out[1]	95	90	85	80	75

(1. For two ½ minute unmuffled blows per 10 hour workday.)

power plant.

UTILITY INDUSTRY ACTIVITIES IN THE UNITED STATES

Electric utility organizations have sponsored several major noise abatement studies that have generated new data which enable engineers to design facilities that emit less environmental noise than those previously constructed. These coordinated efforts and the pooling of resources have resulted in projects that may not have otherwise been undertaken. Concurrent with the efforts of industry groups, individual utilities, using either in-house staff or in conjunction with acoustical consultants and engineers, have also made significant contributions [17, 10, 18, 19, 20, 21, 22 and 23]. Unless there is a significant reversal in national policy towards environmental concerns, efforts to abate noise emissions from electric utility facilities should continue.

Electric utility industry noise abatement activities have focused on quantifying noise emissions from proposed facilities and on abating noise from them. Studies have been sponsored by the Edison Electric Institute, the principal association of investor owned American utilities; Electric Power Research Institute, an organization funded by American electric utilities to develop and manage a technology program for improving electric power facilities; and the Empire State Electric Energy Research Corporation, a research organization sponsored by the New York State utilities [12, 24, 25 and 26].

CONCLUSION

Nonuniformity of noise pollution control regulations, and variations in both equipment sound power levels and site conditions require that noise emissions from each new power plant be uniquely evaluated. An overview of the techniques for performing this evaluation has been presented in this paper. Acceptable community noise levels can be achieved provided that (a) emissions from new facilities are compatible with surrounding land use, (b) sufficient real estate is acquired for a noise buffer, and (c) abatement devices are installed on noise producing equipment.

REFERENCES & SUGGESTED READINGS

1. International Standards Organization, Recommendation 1996, 1971. Available from American National Standards Institute, 1430 Broadway, New York, NY.

10018.

2. Stevens et al, "A Community's Reaction to Noise: Can It Be Forecast?", Noise Control, Vol. 1, No. 1, 1955.

3. F. G. Haag, "Community Response to Industrial Noise," Noise Control Engineering, 3, 1, 10-13, 1974.

4. F. M. Kessler, "Assessment of Acoustic Impact of a Proposed Steam Generating Plant in New York State," Noise Control Engineering, 4, 1, 6-12, 1975.

5. F. M. Kessler, "Acoustic Impact Assessment Procedure Used in Industrial Plant-Site Selection," Noise Control Engineering, 6, 1, 35-39, 1976.

6. R. M. Hoover, "Study of Community Noise Complaints Caused by Electric Power Plant Operations," Noise Control Engineering, 6, 2, 74-80, 1976.

7. US Environmental Protection Agency, Information on Levels of Environmental Noise Requisite to Protect Public Health & Welfare With an Adequate Margin of Safety, Report No. 550/9-74-004, March 1974.

8. A. M. Teplitzky, "Electric Power Plant Noise Emission Controls," Sound & Vibration, September 1976.

9. W. Summer, "Electricity Supply Undertakings & Protection of the Environment," Elektregitatswirt Schaft, November 1973.

10. J. P. Carlson & A. M. Teplitzky, "Estimation and Impact of Environmental Noise from Natural Draft Cooling Towers," Noise Control Engineering, 3, 1, 4-8, 1974.

11. A. M. Teplitzky, "Community Noise Emissions from Enclosed Electric Power Plant," Noise Control Engineering, 6, 1, 4-9, 1976.

12. Edison Electric Institute, Electric Power Plant Environmental Noise Guide. Prepared at the direction of EEI Steering Committee, 1978.

13. A. M. Teplitzky et al., "Electric Power Plant Environmental Noise Guide," Noise Control Engineering, May-June 1981.

14. Ver and Biker, "Design of a Tuned Muffler for Large

Induced-Draft Fans," Proceedings of Inter-Noise '78, 309, 1978.

15. J. P. Reilly, "Power Plant Noise Models for Community Impacts Studies," Proceedings of Noise XPO 1976, 1976.

16. Bolt Beranek & Newman, Inc., Power Plant Construction Noise Guide, Prepared for Empire State Electric Energy Research Corp., May 1977.

17. R. L. Bannister & P. M. Niskode, "Analysis & Control of Steam Turbine Generator Noise," Noise Control Engineering 2, 1, 6-13, 1974.

18. Bartenwerfer, Gikade, Neise, and Agnon, "Noise Reduction in Centrifugal Fans by Means of an Acoustically Lined Casing," Noise Control Engineering, 8, 3, 100-107, 1977.

19. G. Pleeck, "Noise Control in the Turbine Room of a Power Station," Noise Control Engineering, 8, 3, 131-136, 1977.

20. G. Reethof, "Control Valve and Regulator Noise Generator Propagation and Reduction," Noise Control Engineering, 9, 2, 74-85, 1977.

21. J. S. Wang, "Prediction and Control of Induced Draft Cooling Tower Noise," Noise Control Engineering, 12, 2, 74-81, 1979.

22. J. M. Guinter, "Controlling Noise from Electrical Equipment," Noise Control Engineering, 13, 3, 129-144, 1979.

23. A. M. Teplitzky, "Controlling Power Plant Noise," Power, 122, 8, 23-27, 1978.

24. Electric Power Research Institute, Transformer Noise Abatement Using Tuned Sound Enclosures, Report EI-529. Prepared by Allis-Chalmers Corp., 1977.

25. Empire State Electric Energy Research Corp., Characterization of Transformer Noise Emissions. Prepared by Bolt Beranek & Newman, Inc., 1977.

26. Empire State Electric Energy Research Corp., Transformer Noise Abatement, Report EP9-14 Prepared by Westinghouse Electric Corp. & Bolt Beranek & Newman, Inc., 1981.

ENVIRONMENTAL NOISE CONTROL FOR INDUSTRIAL FACILITIES

Robert D. Bruce, Reginald H. Keith,
and Robert M. Hoover

ABSTRACT

In this paper, we discuss the control of environmental problems by industry. First, we present a brief discussion of the criteria used to evaluate environmental noise control problems. Then, we discuss environmental noise control in the power, petroleum, gas transmission, and light manufacturing industries. Finally, we present some recommended guidelines for identifying potential noise problems and correcting them before they become complaints.

*Hoover Keith & Bruce Inc., 11381 Meadowglen, Suite I, Houston, Texas 77082-2632

INTRODUCTION

Industry is concerned about environmental noise and often puts forth extensive effort to control noise emanating from its facilities. Industry's concern about environmental noise existed before the emphasis provided by the Environmental Protection Agency's Office of Noise Abatement and Control which was active in the mid-1970's. Even though that emphasis has been removed, industry continues to provide noise control for selected environmental situations.

In the following section, we present a brief discussion of the criteria used to evaluate environmental noise control problems. Then we discuss environmental noise control in the following industries:

>Power,
>Petroleum,
>Gas transmission, and
>Light manufacturing.

Finally, we present some recommended guidelines for identifying noise problems before they happen.

CRITERIA

Establishing a noise criterion for an industrial site is a challenging endeavor. If there are state or local noise ordinances or, in selected instances, federal regulations that must be met, then these levels represent the maximum values for a potential noise criterion for the site. Even if the facility meets the noise ordinance, this does not ensure that neighbors near the industrial facility will not complain about the noise from the site. Another approach is to use a non-degradation criterion whereby the noise from the industrial facility will not increase the sound level at the property line. This is a very stringent requirement. A third approach is to evaluate the potential for community reaction to the noise. In decreasing order of severity of noise control treatments, these three approaches can be rank ordered, assuming there are nearby residents, as follows:

>Non-degradation,
>Minimum of community complaints, and
>Regulations.

Sometimes the non-degradation criterion is the regulation and all three of these approaches are essentially the same. Our experience suggests that two criterion need to be evaluated--that of minimum of community complaints and regulations, if appropriate. If there is potential for community reaction, then the noise control treatments need to be evaluated for both criterion. Then, corporate management will have the necessary information to decide on a course of action.

There are a number of techniques that can be use to evaluate the potential for community complaint. The approach presented in the Edison Electric Institute's "Electric Power Plant Environmental Noise Guide" <3> is recommended. Basically, this approach evaluates the following information:

- >Noise (level and frequency) content due to plant,
- >Background noise (level and frequency content) without plant,
- >Temporal factors--day vs. night and intermittently,
- >Seasonal factors--winter vs. summer, and
- >Previous exposure by community to industrial sounds.

Using this approach it is possible to estimate the community reaction due to noise from a new plant or addition to an existing facility. If an adverse reaction is predicted, then noise control treatments can be designed into the facility.

POWER INDUSTRY

The power industry has been solving environmental noise problems from industrial facilities for over 35 years. One of the earliest offenders and one which can still cause problems today is noise from induced draft fans. Other power industry sources sometimes causing serious problems are transformers and circulating water pump motors. In addition, forced draft and primary air fans, ventilation fans, boiler feed pumps, safety valves and start-up and blowdown operations can cause community noise problems.

An example of noise control for a power plant is the work performed by New England Power Company at its Brayton Point plant in Somerset, Massachusetts <4>. When Unit 3 was placed in service, noise complaints were received from neighbors about 1000 ft from the plant. When plans were developed for Unit 4, the decision was made to reduce the noise from Unit 3 and to incorporate noise control into the design of Unit 4. A non-degradation criterion was selected.

In order to accomplish this objective, noise control treatments were developed for the following:

SOURCE TREATMENT	NOISE CONTROL
Main Power Transformer for Unit 3	Barrier wall
Forced Draft Fans for Unit 3	Enclosure and muffler
Soot Blower Air Compressor for Unit 3	Exhaust muffler
*Induced draft fans for Unit 3	Mufflers in Discharge
Main Power Transformer for Unit 4	Specified low sound levels
Forced draft fans for Unit 4	Enclosure and intake muffler and duct modifications
Gas recirculation fans for Unit 4	Lagging and barrier wall

*For Coal Conversion

The total cost for these improvements was approximately $2,150,000 in 1981 dollars.

OIL REFINERIES

Petroleum refineries and other plants associated with the petrochemical industry occasionally have community noise problems. Noise from the following sources have caused community noise complaints:

>Electric motors
>Gas turbines
>Furnaces
>Flares
>Air-cooled heat exchangers
>Cooling towers
>Boiler feed pumps
>Blowdowns
>Large fans and blowers

Electric motors

Occasionally totally-enclosed fan cooled motors without noise control treatments have caused noise problems in communities. Many of these motors were cooled with flat-bladed radial fans that can be operated in both directions; this flexibility resulted in a noisy fan design. Today, one can purchase quiet designed motors that produce sound levels less than 85 dB(A) at 3 ft.

Gas turbines

Gas turbine noise can be troublesome in the community, particularly the low frequency noise from the exhaust.<5> In addition to the low frequency exhaust noise, high frequency noise is radiated out the intake and broad band noise is radiated from the casing. Also, the driven component, such as a generator or compressor, may also require treatment. Noise control is accomplished by enclosing the gas turbine and providing mufflers for the intake and the exhaust. In addition, some work has been done with active sound cancellation of low frequency exhaust noise but this work is still in the development stage.<6> Gas turbine noise control is well understood and it is possible to purchase new gas turbines that meet noise criteria.

Furnaces

Furnace noise is due to the combustion of the fuel and to the inspiration of air for combustion. In addition to direct radiation from the burner, it is also possible to have radiation from the walls of the furnace. The burner noise can be controlled by providing an intake air plenum around the burners. Radiation from the walls of the furnace is a more difficult problem, requiring individual analysis and treatment.

Flares

Flares produce noise by combustion of the excess hydrocarbons. In addition, the steam that is injected into the flame to eliminate the smoke and the luminosity of the combustion increase the noise. The steam noise can be controlled through the use of multi-port nozzles. Alternatively, in selected instances, a ground flare can be used. The ground flare can be quieted through the use of enclosure techniques. <1>

Air-cooled heat exchangers

Noise of air-cooled heat exchangers or fin-fans, as they are sometimes called in the industry, is due to the turbulence created by the blade passage. Noise control for fin-fans is very difficult after the fans have been installed. Attention should be given to purchasing quieter units.

Cooling towers

Cooling tower noise control solutions have been available for many years. However, we occasionally see an installation where the cooling tower is a problem. Barriers and mufflers are typical treatments. Again the purchase specification can be used to ensure the installation of a quieter unit.

Boiler feed pumps

Large boiler feed pumps can sometimes be a community annoyance problem. These problems involve the vane-passage frequency of the pump and are often solved by modifying the cut-off or the vane or by lagging the pumps with acoustical insulation or enclosing them.

Blowdowns

The noise of venting steam to the atmosphere is often caused by turbulent mixing at the exit plane of the stack or by the valve in the blowdown line. These problems can be solved through the use of a silencer or a quieter valve.

Large fans and blowers

Fans and blowers can be community noise problems. Most often the problem is due to the blade passage tone of the fan. This is particularly true for radial bladed centrifugal fans and vane-axial fans. It is also true to a lesser degree for other types of centrifugal and axial fans. Community noise problems are usually the result of the discharge noise of induced draft fans or primary air fans. Noise control usually consists of a muffler treatment, either in the discharge breeching or in the stack. The mufflers can be of the dissipative type, the reactive type or a combination. In addition to mufflers, in some instances, modifying the fan cut-off or increasing cut-off clearance has lead to some success. Other measures which are under some development are the use of resonators at the fan cut-off and the use of active cancellation of sound.

Rotary-positive displacement blowers are often problems as well and can be treated with mufflers.

Additional information about noise sources in the petrochemical industry can be found in Reference 2.

GAS TRANSMISSION

Currently natural gas pipeline companies transporting gas interstate operate their compressor stations under the auspices of the Federal Energy Regulatory Commission (FERC). Under Section 7 of the Natural Gas Act, interstate pipeline companies can apply for a blanket certificate for operation of a compressor station.<7> One of the provisions stipulates that the noise due to the compressor facility can not exceed a day-night sound level (Ldn) of 55 dB(A) at nearby noise-sensitive areas, such as schools, hospitals, and residences. Since the Ldn value penalizes night-time operations and since compressor stations operate 24 hours per day, the continuous noise level represented by an Ldn of 55 dB(A) is about 48 dB(A). The regulatory limit is not dependent on the existing noise levels in the area. In other words, if the existing levels are greater than 55 dB(A), the compressor station is still limited to 55 dB(A). Thus, the compressor station is not penalized for the existing ambient. If the existing levels are less than 55 dB(A), then the levels due to the compressor are still limited to 55 dB(A). At first glance, it appears that compliance with the noise limits should ensure that the acoustical problems related to the station will be solved. Unfortunately, it is possible that compressor stations can be in compliance with the noise limits and still generate sufficient annoyance that neighbors will complain. This is particularly true in rural areas where there has been no major noise source before the compressor station went on stream.

It is not uncommon for these compressor stations to be built within 1000 ft of residences. The compressor stations vary in size but they will typically have the following equipment:

>Several diesel engines (typically 8 or 12 cylinder units) or gas turbines,
>Inlet blowers on diesel engines,
>Several compressors,
>Cooling fans for engines,
>Valves and piping,
>Gas coolers, and

>Blowdown lines.

Although most diesel engines will be installed with exhaust mufflers, the mufflers are often not selected such that they will meet the noise requirements. In addition to the engine exhaust, each of the above noise sources has been the source of noise complaints in the gas transmission industry in the past. Noise control treatments range from constructing a building around the entire operation to building absorptive barrier walls between the sources and the nearby residents. In addition, it has been necessary to:

>Install larger mufflers on the engine exhaust,
>Reduce speed on compressors,
>Remove turbo-chargers from engine intakes,
>Install mufflers on gas coolers,
>Wrap control valves, and
>Provide blowdown silencers.

The noise problems of the gas transmission industry can be evaluated and solved before the system goes on stream. In order to accomplish this, it will be necessary to consider the acoustical problems in the design stage and to use the purchase order to require vendors to provide equipment designed and manufactured to limit the noise emissions.

OTHER INDUSTRIES

Most light manufacturing industries near residential areas have the potential for noise complaints. Many of these complaints are due to noise from fans and other air moving systems. In addition, trash, delivery, and refrigeration trucks, transformers, public address and other communication systems, and relief valves can generate noise levels that are annoying to nearby residents.

All of the environmental noise problems faced by light industry can be solved in the design stage if someone recognizes the potential for a problem.

GUIDELINES FOR FUTURE INSTALLATIONS

It is possible to design noise control treatments into an industrial facility so that the noise from the facility does not violate the noise limits set by the pertinent regulatory authority. Alternatively, it is possible to establish noise criteria for an industrial facility such that complaints from neighbors about noise are minimized. The first step is to decide which of these two criteria are to be used in evaluating the noise problem. Even if the decision is made to design only to the regulatory requirements, we recommend that the amount of noise reduction that is required in order to minimize neighbor complaints be determined.

Alternative noise control treatments can be developed to provide the necessary noise reduction. These treatments can be evaluated from the perspective of operations, maintenance, and cost. The optimum system can be selected. One very effective approach is to require the vendor to provide equipment that has been designed to be quieter than the standard unit. This requires planning on the part of the User as well as the Original Equipment Manufacturer of the equipment item. This approach has been effective in purchasing power transformers, electric motors, and compressors, for example.

SUMMARY

In this paper, we have reviewed several criteria for use with environmental noise evaluations. Then we have given examples of the types of equipment in the power, petroleum, gas transmission, and light manufacturing industries that have given rise to noise complaints. In addition, we have identified some of the noise control treatments that have been used to reduce the noise.

REFERENCES

1. Bruce, R. D., "Noise Control for a Flare Near a Community," 38th Petroleum Mechanical Engineering Workshop and Conference, Sept. 12-14, Pennsylvania, 1982.

2. Bruce, R. D., "NOISE POLLUTION," KIRK-OTHMER: ENCYCLOPEDIA OF CHEMICAL TECHNOLOGY, Volume 16, Third Edition, John Wiley & Sons, Inc., 1981.

3. Edison Electric Institute, "Electric Power Plant Environmental Noise Guide," available from EEI, 1111 19th St., N.W., Washington, D.C. 20036.

4. Hoover, R. M., "Noise Control at Fossil Fuel Power Plants: An Industry-wide Assessment of Costs and Benefits," EPRI CS-3262, Final Report, Dec. 1983.

5. Hoover, R. M., "The Sound of Gas Turbine Installations," ASME Publication 70-WA/GT-6 paper presented at the ASME Winter Annual Meeting, New York, Nov. 29-Dec. 3, 1970.

6. "Killing noise with noise: a sound solution?," Electrical Review Vol. 211 No. 17, 19 Nov. 1982, pp. 22-23.

7. Natural Gas Act, Sub-Part F of Part 157, Paragraph 157.206(d)(5).

Cost-Effective Design of Practically Vibration-Free High-Technology Facilities

Eric E. Ungar* and Colin G. Gordon**

Abstract and Introduction

Advanced electro-optical equipment, such as that used in the production of integrated circuits, requires clean environments with extremely limited vibrations. Ground motions, personnel activities, and mechanical equipment tend to produce unacceptably severe vibrations in high-technology facilities, unless suitable precautions are taken.

The first of the following sections deals with recommended vibration criteria and the second presents an overview of the important vibration sources. Subsequent sections provide general guidance and suggests means for control of vibrations produced by external sources, interior activities, and mechanical equipment.

Vibration Criteria

A completely vibration-free environment is as unachievable as are such other idealized abstractions as immovable objects, irresistible forces or perfect vacua. Fortunately, it is generally sufficient in practice to provide a vibration environment that is adequately vibration-free - that is, an environment that does not exceed suitably selected vibration limits.

For high-technology facilities - for example, such as are used for fabrication of integrated circuits - limits on the permissible vibrations logically are set by the most severe vibration environments under which critical items of vibration-sensitive equipment can operate satisfactorily. If all equipment items to be placed in a facility under design are fully identified, and if the acceptable vibration limits of all items are known, one may readily derive a vibration criterion for the facility, e.g., by requiring the environmental vibration in each frequency band not to exceed the vibrations that are acceptable for the equipment with the most stringent limitation in that band. (Note that different equipment items may determine the criterion values in different frequency bands.)

---

*Chief Consulting Engineer, BBN Laboratories, 10 Moulton St., Cambridge, MA 02238

**Manager, Los Angeles Office, BBN Laboratories, 21110 Vanowen St., Canoga Park, CA 91305

However, establishment of realitic vibration criteria for a facility usually involves some complications. Quite often, the equipment to be installed in a facility is not fully known at the time the facility is being designed - and, more importantly, the installed equipment may be changed in the future, as new processes and equipment become available. Furthermore, even for some items of sensitive equipment that are presently in wide use, the acceptable vibration limits are not known adequately. Although most equipment manufacturers provide some vibration specifications, many such specifications are based only on estimates and tend to be overconservative, many do not indicate the range of frequencies within which a specified limitation applies, and virtually none identify the frequency bandwidths in which measurements are to be made.

In order to develop meaningful facility environmental vibration criteria in light of the aforementioned difficulties, we have reviewed a considerable number of equipment criteria supplied by equipment manufacturers and also have made exploratory measurements on several different items of equipment. This work has led us to develop the criteria shown in Fig. 1, which we have used extensively for design and evaluation purposes.

The frequency range associated with the criteria of Fig. 1 takes account of two important considerations. The lower end of the frequency range is based on the fact that the operation of optical or mechanical devices is affected by internal (relative) deflections, and not overall (absolute) deflections - and that base motions at frequencies that are well below a device's fundamental natural frequency produce only very small relative deflections. The higher end of the frequency range is set by the fact that the vibration isolation systems, as well as the structural components of the sensitive equipment items themselves, transmit less vibration at higher frequencies.

The vibration criteria indicated in Fig. 1 are expressed in terms of vibrational velocity, even though vibrations generally are measured most directly in terms of acceleration, and even though vibrational displacements can be visualized more readily. Of course the amplitudes of displacement x, velocity v, and acceleration a, at any frequency f, are interrelated simply by

$$a = 2\pi f v = (2\pi f)^2 x,$$

so that one readily can convert from one representation to another. Our use of velocity is not based on a fundamental principle, but on our observations that the many relative mimima of the typically wildly fluctuating curves which represent the frequency-variations of the measured maximum acceptable vibration amplitudes for items of sensitive equipment generally lie approximately on a curve of constant velocity. Such a curve constitutes a simple conservative vibration criterion which has the advantage that it can be characterized essentially by a single number (without a frequency dependence) - namely, the maximum acceptable velocity - against which measured or predicted vibrations can be judged.

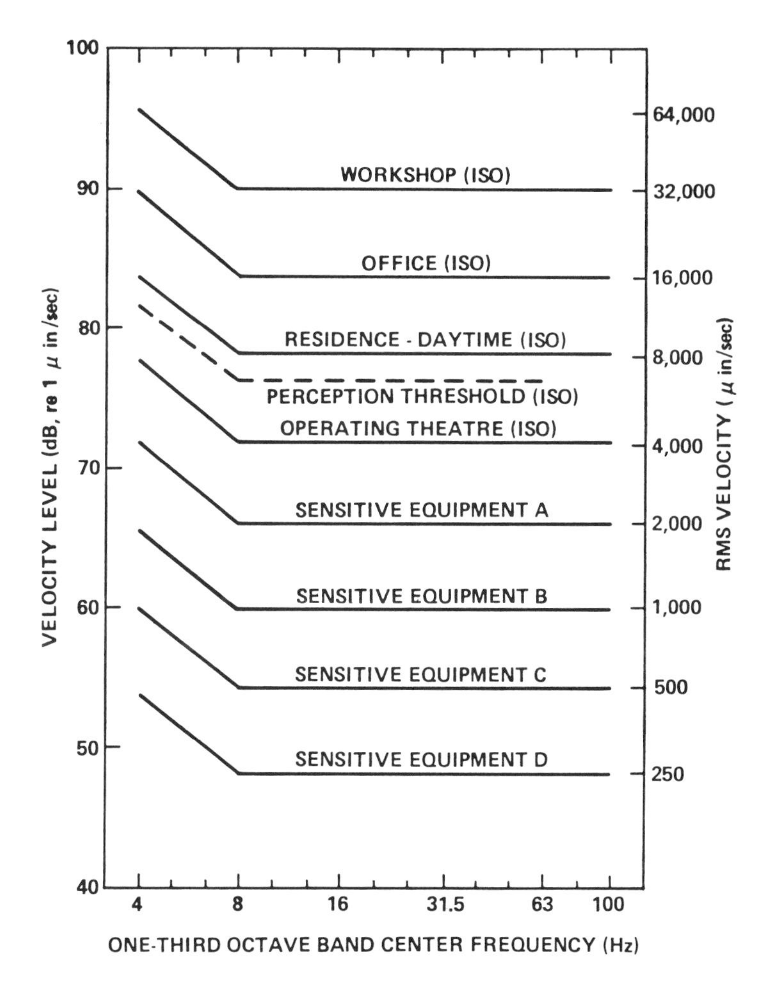

Fig. 1 Vibration Criteria for Sensitive Equipment in Buildings

A: Optical balances, bench microscopes
B: Aligners, steppers, etc. for 5 μm or larger geometries
C: Aligners, steppers, etc. for 1 μm or larger geometries
D: E-beam and other 1 μm or sub-micron equipment; electron microscopes

The "Sensitive Equipment" curves of Fig. 1 represent, in our view, the requirements for the most vibration-sensitive equipment items in each category. (They thus are more restrictive than necessary for other items in the category that are less sensitive to vibration.) Criterion D represents a conservative requirement for electron-beam equipment and for scanning electron microscopes at magnifications up to about 2000x (as normally used in the microelectronics industry); a more stringent requirement may apply to some electron microscopes at their upper limit of magnification. Criterion C represents, in our opinion, a suitable design standard for current state-of-the-art microelectronics manufacturing facilities, and may also be quite adequate for future sub-micron microelectronics technology, given some selectivity in the process equipment to be used.

In addition to the "Sensitive Equipment" criteria curves, Fig. 1 shows, for the sake of comparison, several curves indicating vibration limits in relation to human comfort, as well as the whole-body threshold of perception set forth by the International Standards Organization (Ref. 1). One may note that sensitive equipment requires considerably tighter limits on vibration than even the most sensitive activities of people.

Vibration Sources

The major sources of vibrations of concern in relation to high-technology facility buildings fall into three categories: (1) external sources, (2) internal activities and (3) service machinery.

External sources include ambient vibrations at the site (sometimes called micro-tremors), nearby road and rail traffic (including underground and elevated roads and rail systems), construction activities (including blasting), and machinery operating in the vicinity (either outdoors or in nearby buildings).

Internal activities include personnel walking (footfalls) and service activities (e.g., repair and construction) in-plant vehicles (such as forklifts and carts), and production work (e.g., actuation of production machines or other tools).

Service machinery includes all mechanical and electrical equipment that services the building and supports the production processes. It includes air-conditioning and distribution fans, chillers, cooling towers, furnaces, and all pumps, compressors and vacuum pumps, as well as elevators and mechanically actuated doors and loading platforms.

Dealing with External Sources

It stands to reason that extremely vibration-sensitive facilities should be located in areas where ambient vibrations in the ground are acceptably small, where there exists no significant nearby road or rail traffic, and where one expects no continuing construction activity or other heavy machinery operation in the vicinity. A site vibration survey is essential for evaluating potential facility sites in this regard. In such a survey, careful attention needs to be paid to such considerations as the seasonal variations in vibration-transmitting properties of the ground (which may depend on the moisture content of the soil and the height of the water table, among other things) and the daily variations in traffic.

At a given building site, the vibration-sensitive activities generally should be located as far from external vibration sources as possible, in order to take advantage of the vibration attenuation that usually is associated with increased distance. However, this attenuation may not be realized if the soil at the site is not uniform; a detailed vibration survey can define the vibration distribution at the site and enable placement of the most sensitive operations in the most vibration-free locations.

In some instances, one may be able directly to reduce the vibrations generated by an external source. For example, it is well known that the most severe vibrations associated with road traffic result from heavy vehicles with stiff suspensions moving rapidly along roads with surface irregularities. Thus, one may keep heavy trucks away from sensitive facilities, one may limit the permissible speeds, and one may smoothe the road surface. Certainly, "speed bumps," potholes, misaligned slabs and expansion joints (in bridges) should not be permitted near vibration-sensitive facilities. Similary, if the related costs are acceptable, one may consider replacing jointed rail by continuously welded rail in rail lines passing near such facilities, and/or placing such rail lines on thick ballast beds or on resilient rail support systems (e.g., Ref. 4).

We are aware of no practical means for attenuating vibrations that propagate along the ground. Berms, heavy walls, and other structures above the ground have very little effect on waves at the frequencies of primary concern here. The same is true of trenches, sheet piling and slurry walls and similar underground structures or geotechnical means (e.g., grout injection) of practical size.

However, some benefit can be obtained from appropriate foundation design and from isolating key parts of the facility from soil vibrations. In situations where the ambient vibrations of the bedrock are of relatively small magnitute, compared to those of the surface soil, it is advisable to base the building foundations on the bedrock and to avoid their coupling to the surface soil. On the other hand, where the surface soil vibrates less than the bedrock, mat foundations or spread footings are preferable. Which of these situations exists depends on the soil conditions and on the locations and types of the predominant external vibration sources. By choosing appropriate footing designs, one also may make the soil and footing behave like a suitably tuned spring, acting in conjunction with the mass it supports to attenuate transmitted vibrations above a selected frequency, much as does a classical mechanical spring-mass system. The frequency-distribution of the ground vibration must be taken into account here, however, because such spring-mass systems also amplify vibrations in the vicinity of the systems' resonance frequencies.

Where footing design cannot provide sufficient attentuation, selected parts of the building may be isolated from ground vibrations by supporting them on resilient elements, such as neoprene bride-bearing pads or "air mounts" (pneumatic springs) (Ref. 2). Here again, these resilient elements and the mass they support act like a mass-spring system, which attenuates vibrations above its resonance frequency, but amplifies vibrations near that frequency. Such a specially designed system has the advantage that it can be made very resilient, with a resonance frequency below the range of concern, and with the potential for providing considerable attenuation of intruding groundborne vibrations. It has the disadvantages of considerable complexity and attendant costs, as well as the potential for increasing vibrations resulting from internal activities.

Reduction of intruding vibrations in selected frequency ranges may also be achieved by mounting resiliently supported masses to the foundations, so as to produce in essence in a "tuned absorber" system. Such an arrangement, which has been used successfully in buildings in Japan in a different context, (Ref. 5), tends to require considerable mass and careful design.

Control of Vibrations due to Internal Activities

The problem of vibrations caused by footfalls (walking personnel) needs to be addressed early in the facility design process, because it generally requires a structural or architectural solution. Footfall-induced vibrations tend to be relatively insignificant for slabs on grade, but usually are of major importance for above-grade floors.

An above-grade floor may be visualized as acting somewhat like a trampoline; footfall impacts on it set it into motion, and thus also set into motion any equipment resting on it. Footfalls near the center of a bay tend to cause the greatest vibrations, and the vibrations always tend to be most severe at mid-bay and least severe near columns. Thus, footfall-induced vibrations and their effects may be reduced by confining heavily travelled areas (e.g., visitors corridors) to regions near column lines, placing sensitive equipment near columns, and keeping as much distance as possible between heavily travelled areas and sensitive equipment.

It is also well-known that rapid walking causes more severe footfall impacts than slower walking and that several people walking in step can cause very strong vibrations. The probability of obtaining such conditions may be reduced by instituting administrative controls or by avoiding long straight corridors that permit rapid walking.

However, the most reliable solutions to the footfall-induced vibration problem usually are structural, consisting of (1) separating the structures on which people walk from those that support the sensitive equipment, and (2) making the floor structures stiff and massive enough so that the footfall-induced vibrations remain within acceptable limits. The structural separations may be horizontal (e.g., in the form of resilient joints between corridors and sensitive fabrication areas) or vertical (e.g., where personnel are confined to a "bridge" that is supported from the columns without making contact with the process floor.)

Analytical methods for predicting the magnitudes and spatial distributions of footfall-induced vibrations have been developed and validated (Ref. 6); they are convenient for evaluating specific structural floor designs and for indicating what changes are necessary to achieve desired vibration criteria. Any type of structure - steel, concrete, or composite - can be designed to perform adequately from this standpoint, but concrete waffle-slabs have been used most widely. Usually the flexural stiffness of the floor structure turns out to be the controlling parameter, and the greatest flexural stiffnesses generally can be obtained by keeping the column spacings as small as possible.

The aforementioned concepts for controlling footfall-induced vibrations also are useful for limiting the vibrations due to in-plant vehicles. However, a vehicle entering on a floor slab or leaving a slab also in effect produces a suddenly applied load; it is desirable to reduce the suddenness of load application - e.g., by using joints with long interlacing fingers or by having joints arranged so that only one wheel of a vehicle crosses the joint at a time. It is also useful to use soft pneumatic tires on all vehicles and keep the roadways smooth and free of surface discontinuities.

The effects of vibration-producing production-related machines can be reduced by keeping these as far from sensitive equipment areas as possible, by locating these machines in areas where the supporting structures are relatively stiff (e.g, near columns), and by supporting these machines on resilient vibration-isolating systems. In general, the same vibration control concepts that are discussed in the following paragraphs in relation to service machinery apply here also.

Control of Vibration Produced by Machinery

In the early stages of design of a facility, there often exist opportunities for selecting mechanical and electrical equipment types that inherently are relatively free of vibration. For example, rotating compressors tend to produce considerably less severe vibrations than reciprocating compressors, because their inertia forces are better balanced; for the same reason, multi-cylinder (particularly opposed-piston) engines and compressors are preferable to single-cylinder machines. Similarly, it is advisable to choose the better balanced of two otherwise similar machine models, and one may do well to opt for the purchase of equipment with the best economically feasible field-balance specifications.

It is advisable to keep as much distance between vibration-sensitive equipment and vibration-producing machinery as possible, to support vibration-producing machinery on stiff structural components, and to provide this machinery with efficient vibration isolation systems.

For the purpose of selecting machinery balance specifications and isolation systems, one may do well to take note of the following easily derived expression for the velocity amplitude V of the point on a structure that supports a piece of vibration-producing machinery, keeping in mind that the vibratory velocity at other locations in the building is proportinal to that at the support point:

$$V = (2\pi N)X$$

$$X = \frac{W}{k_o}\frac{x_u}{\delta_{st}} = \delta_o \frac{x_u}{\delta_{st}} .$$

Here x represents the displacement amplitude of the support point and N the rotational speed of the machine (i.e., of the shaft associated with the machine unbalance responsible for the vibration). The unbalance of a machine typically is measured by mounting the machine on a system of very soft springs, running the machine, and measuring the resulting vibratory displacement (or excursion) of its base; this "unbalance displacement" amplitude is designated by x_u. The symbol W represents the total weight of the machine (including the weight of any base on which it may be mounted) and δ_{st} denotes the static deflection of the resilient isolators on which the machine is to be supported in the actual field installation; that is, δ_{st} represents the deflection of the isolators that results from application of the weight W. Finally, k_o represents the stiffness of the supporting structure, and δ_o denotes the supporting structure's static deflection caused by the weight W.

From the foregoing expression, one may observe that the velocity amplitude is proportional to the unbalance amplitude x_u and to the rotational speed N, as well as to the ratio δ_o/δ_{st} of the static deflection of supporting structure to that of the isolation system (which ratio also is proportional to the ratio of the stiffness to the isolator support structure stiffness). One may thus readily deduce the importance not only of using soft resilient isolators, but also of supporting the machines on stiff structures.

The often-heard statement "higher-speed machinery is preferable because it can be isolated better" is not generally true in the present context. (This statement is based on a classical textbook analysis applicable to cases where the magnitude of the force acting on a machine mass does not change with rotational speed, whereas for a given rotating machine the unbalanced inertia force varies as the square of rotational speed.) As is evident from the foregoing expression, the structural vibration velocity V varies as both N and x_u, so that for a given amount of unbalance a higher speed machine may be expected to produce more severe vibrations than a similarly isolated low-speed machine.

It thus is inappropriate to conclude that centrifugal fans installed in a plant constitute more signficant vibration sources than vane-axial fans, because the latter rotate faster and therefore can be isolated better. Although high-quality vane-axial fans typically can be balanced better than well-balanced centrifugal fans, it often turns out that the values of the Nx_u product for competing fan systems are very similar. Because the achievable isolator static deflection is virtually independent of the system being isolated, the selection of fans generally needs to be made on the basis of a comparison of the Nx_u product values for candidate systems.

The addition of an inertia base (i.e., a large mass) to a vibrating machine increases the total weight W while reducing the excursion x_u proportionately, keeping the product Wx_u unchanged. One may conclude from the previously given equations that an inertia base has no effect on the vibration transmitted to the structure, provided that the static deflection of the isolation system is kept unchanged. (This is usually the case, because one typically tries to use the softest practical isolation system.) On the other hand, addition of an inertia base does decrease the vibratory excursion of the isolated machine, thus reducing the vibrations that are induced in attached piping, ducts or conduits.

Ancillary Considerations

In high-technology facilities, as in any complex dynamic system, careful attention needs to be paid to a large number of details, in order to ensure that the desired vibration performance is indeed obtained. All potential vibration transmission paths that may "short circuit" machinery isolation systems or structural breaks need to be considered and eventually treated. This includes piping, ducts, conduits that may bridge the isolation systems or gaps, as well as such auxiliary structures as partitions and pipe racks.

Good design from the start is important, but even the best design is useless unless it is implemented properly. For this reason, we typically advocate not only careful review of relevant shop drawings, but also repeated field inspection in the course of construction. Ideally, vibration measurements should also be performed after the facility is completed, so that meeting of the vibration specifications can be verified and any residual problems can be identified and resolved.

It should also be noted that audible noise tends to be a significant problem in high-technology facilities that involve "clean room" installations. These clean rooms require large amounts of air, and the airhandling systems that supply this air are sources of both noise and vibration. It usually is advangageous to address the noise contol considerations simultaneously with vibration control, so that compact, cost-effective treatments can be devised.

Concluding Remarks

Although the attempt was made in the present paper to summarize the most important factors in the design of practically vibration-free facilities, it obviously does not treat the multitude of significant details that must be considered in practice. Proper consideration of all vibration aspects generally requires the services of specialists who are familiar both with the underlying theory and with the practical hardware. Ideally, such specialist should collaborate with the facility planners, users, architects, structural engineers and mechanical engineers from project site selection through final checkout.

REFERENCES

1. Anon. "Guide for the Evaluation of Human Exposure to Whole-Body Vibration." ISO Standard (Draft) 2631; 1974.

2. Anon. "Sperry IC plant floats on concrete and airbags to prevent vibrations." Electronics, April 19, 1984; pp. 54-55.

3. Gordon, C.G. and E.E. Ungar "Vibration as a Parameter in the Design of Microelectronics Facilities." Transactions, Inter-noise '83, 1983, pp 483-486.

4. Kurzweil, L.G. and E.E. Ungar, "Prediction of Noise and Vibration in Buildings Near the New York City Subway." Transactions, Inter-Noise 82, 1982, pp. 213-216.

5. Konagai, K., K. Shimizu, N. Sano and T. Ikeda. "A Method for Reducing the Ground Tremor and Building Vibration Causes by Exciting Forces on a Tunnel Floor." Transactions, Inter-noise '84, 1984, pp 559-562.

6. Ungar, E.E. and R.W. White "Footfall-Induced Vibrations of Floors Supporting Sensitive Equipment." Sound and Vibration, October 1979, 13, pp 10-13.

7. Ungar, E.E. and C.G. Gordon "Vibration Challenges in Microelectronics Manufacturing." Shock and Vibration Bulletin 52, Part 1, May 1983, pp 51-58.

8. Ungar,E.E. and C.G. Gordon "Vibration Criteria for Microelectronics Manufacturing Equipment." Transactions, Inter-noise '83, 1983, pp 487-490.

Site Selection and Building Design for Minimizing Vibration

James T. Nelson,* Warren E. Blazier,** Hugh J. Saurenman*

Over the last five to ten years, line resolutions required for state-of-the-art semi-conductor wafers have dropped to well below 1 micron. As a result, controlling environmental building vibration caused by sources both within and without a production facility has become very important. A number of steps can be taken to ensure that the vibration environment will be satisfactory for vibration sensitive equipment. These include (1) performing ground vibration surveys to aid site selection for a new facility; (2) performing vibration surveys within existing buildings to determine appropriate locations for vibration sensitive equipment and processes; (3) reviewing vibration specifications for sensitive equipment; (4) providing special vibration control measures in the building foundation and structural design; and (5) vibration isolation of building mechanical equipment. With careful attention to the vibration environment and specifications for vibration sensitive equipment, vibration problems in the semi-conductor industry can usually be avoided.

Introduction

Over the past ten years, vibration at manufacturing facilities for large scale integrated circuits has become a primary consideration in site selection and building design. The semi-conductor industry is pushing optical resolutions to perhaps 0.25 microns in order to develop new products and technologies. Vibration affects the actual fabrication of semi-conductor chips as well as computerized wafer and reticle inspection machines. Since a significant fraction of state-of-the-art semi-conductor chips may be rejected due to flaws, controlling vibration as well as dust and temperature is extremely important.

Our experience in the semi-conductor industry is that if proper attention is directed to both the siting and design of buildings as well as in the design of production and inspection equipment, vibration can be effectively controlled to the point of insignificance. Where we have encountered a vibration problem, we usually find that the floor on which the equipment is mounted usually has an inadequate stiffness, resulting in a 1st mode

*Wilson, Ihrig & Associates, Inc., 5776 Broadway, Oakland, CA 94618
**Warren Blazier Associates, Inc., Suite 2840, Four Embarcadero Center, San Francisco, CA 94111

resonance frequency lying within a range that cannot be tolerated by the equipment itself. Therefore, it is essential that the building designer be advised of the vibration sensitivity of the equipment (both amplitude and frequency) before the structural design and configuration of the facility is finalized.

The subjects discussed in this paper concern methods for evaluating potential sites for vibration sensitive processes, vibration criteria for buildings, and building design considerations. The intent is to provide an overview and suggest approaches which may be taken when considering potential site locations for manufacturing facilities or solving existing vibration problems. The discussion applies equally to any laboratory facility where sensitive processes are performed. Examples of these processes include growing synthetic crystals, optical cell and chromosome sorting, and electron microscopy. Vibration control will only become more important as the state-of-the-art regarding these processes is advanced.

Vibration Criteria

Vibration criteria must be established for siting vibration sensitive processes. These criteria must be based on a knowledge of the vibration tolerance of the process equipment and the limiting constraints imposed by practical and cost-effective building design considerations. At the present time, the tolerance of various types of equipment to vibration is not generally known, and, as a result, conservative building design approaches are being employed for new facilities.

Some manufacturers of vibration sensitive equipment have recently begun to specify maximum floor vibration levels for proper operation of their equipment. Since this equipment may cost upwards of $500,000 per unit, a considerable investment is at stake in guaranteeing proper operation.

Recently, we have been involved in determining the vibration tolerance of certain items of equipment used in the semi-conductor industry. The procedure involves the use of an electromagnetic shaker to excite the equipment under test with a slowly sweeping sinusoid spanning the frequency range of about 1 Hz to 100 Hz. The amplitude of the excitation is slowly increased until a malfunction is detected and then decreased until a satisfactory operating condition is just obtained. An example of a vertical vibration tolerance curve measured by this technique is presented in Figure 1 in terms of the vibration acceleration level in dB. The equipment under test was an optical inspection machine used in the semi-conductor manufacturing industry.

The minima of the tolerance curve define a maximum sinusoidal vibration level which the equipment can withstand without malfunctioning. Specifying this maximum level, reduced by some factor of safety, thus establishes a practical criterion for siting the machine. Our experience, however, is that these vibration limits should be presented primarily in terms of vibration velocity, rather than acceleration as was done in Figure 1.

Below the lowest resonance frequency of about 2.6 Hz for the equipment considered in Figure 1, or lowest frequency at which a dip in the tolerance curve occurs, the tolerance to vibration acceleration is constant with decreasing frequency. This is the "stiffness controlled" region. Some "specifications" indicate exactly the opposite, namely a constant displacement curve for the stiffness controlled region, primarily (we suspect) because of limitations in the shakers used to test the equipment. Thus, many specifications may be unrealistic and unnecessarily restrictive at low frequencies. A happy medium is simply to specify a maximum vibration velocity over a bandwidth of, for example, 2 Hz to 100 Hz. Judging from the data given in Figure 1, an appropriate floor vibration velocity level for this particular machine is about 1000 micro inches per second.

Our experience suggests that when floor vibration velocity levels exceed about 50 dB re 10^{-6} in/sec., or 300 micro inches/second, we may begin to expect problems with poorly designed or very sensitive optical inspection equipment. At 60 dB, or 1000 micro inches/second, vibration levels are usually excessive for most sensitive equipment. Below 40 dB, or 100 micro inches/second, vibration is usually not a problem. Thus, a maximum floor vibration criterion of about 200 micro inches/second (46 dB re 10^{-6} in/sec) is generally appropriate for most semi-conductor manufacturing processes. This level of floor vibration can usually be obtained with practical building designs.

Vibration Surveys

Vibration surveys should be performed to determine acceptability of potential construction sites for new buildings, or at potential locations for sensitive equipment within existing buildings. These surveys should be statistical in nature and provide data which the building designer or building engineer may use for recommending vibration control provisions.

A vibration survey consists of recording ground and/or building vibration data over perhaps 10 minute periods during selected daytime and nighttime periods. The following time periods are usually appropriate for ground vibration surveys at candidate building sites:

Daytime	10 am	to	2 pm
Rush hour	3 pm	to	5 pm
Evening	7 pm	to	10 pm
Night	11 pm	to	2 am

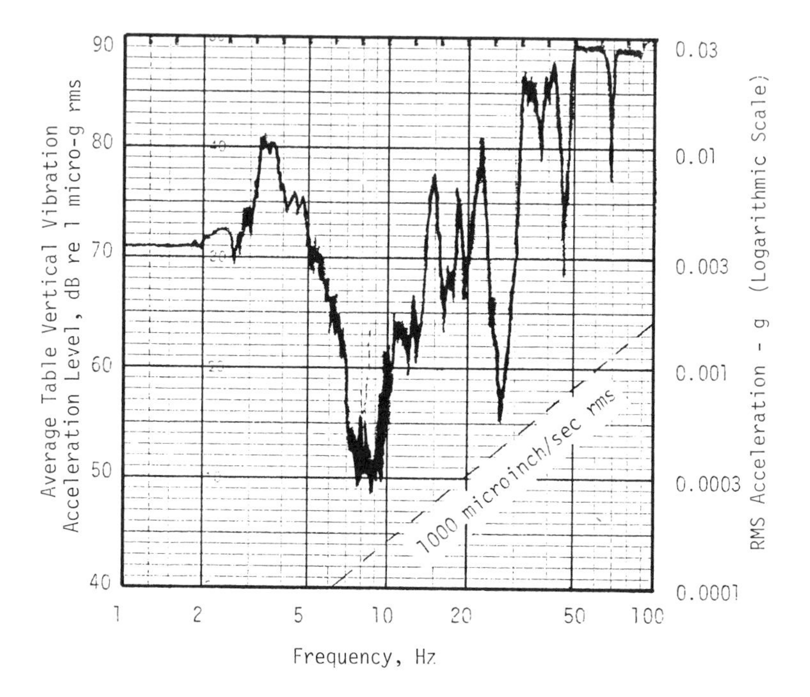

Figure 1. Example of the Threshold of Failure Due to Sinusoidal Vertical Vibration

Our experience indicates that samples taken within these time periods are representative of the vibration environment due to community sources such as traffic. If plant operations are not anticipated during evening or nighttime hours, then these time periods may be ignored. On the other hand, if an excessive vibration level over a short period of time can destroy or interfere with a process which may require several hours or even days to complete, longer sample periods or possibly continuous 24-hour surveys are appropriate.

In some cases, full triaxial vibration measurements are appropriate. We usually limit survey measurements to the vertical direction only, since this is usually the component with the highest level. This is especially true for ground vibration propagated as Rayleigh surface waves and for floor vibration due to bending. Horizontal vibration data can be important, however, for evaluating sites for equipment which exhibit a low tolerance to primarily horizontal vibration, or sites where extremely critical processes are to occur.

For many of the surveys we have performed, the recorded vibration velocity data are analyzed statistically using a 1/3 octave real-time analyzer and computer to produce histograms for each 1/3 octave. An example of this type of analysis is presented in Figure 2. Spectra are plotted for the levels L_1, L_{10}, L_{50} and so forth, where L_n is the vibration velocity level exceeded N- percent of the time. Also plotted is the equivalent continuous level, L_{eq}, or rms vibration spectrum measured over the sample period. The data provide not only a statistical description for a site, but also spectral information which can be used to determine appropriate vibration control procedures.

In addition to statistical analyses, the survey data should contain spectra for identifiable transient events such as railroad train passage and auto and truck traffic. In the case of existing buildings, vibrations produced by individual building mechanical equipment such as HVAC systems, production equipment, and people walking should be documented. Vibration due to footsteps, doors opening and closing, carts, etc. are usually most significant, especially for poor floor constructions. Footfall induced floor vibration is discussed in detail by Ungar and White (Reference 4).

Transfer mobility measurements can be considered at proposed construction sites to assess the capability of the ground to propagate vibration. We have not as yet performed such measurements for proposed semi-conductor production sites, but we have performed such tests for the purpose of predicting ground vibration produced by transit trains. We have found the technique to be very powerful in separating ground response characteristics from source characteristics. The method could be extended to predicting ground vibration caused by punch presses, stamping mills, or other heavy equipment, for a specific site (Reference 3).

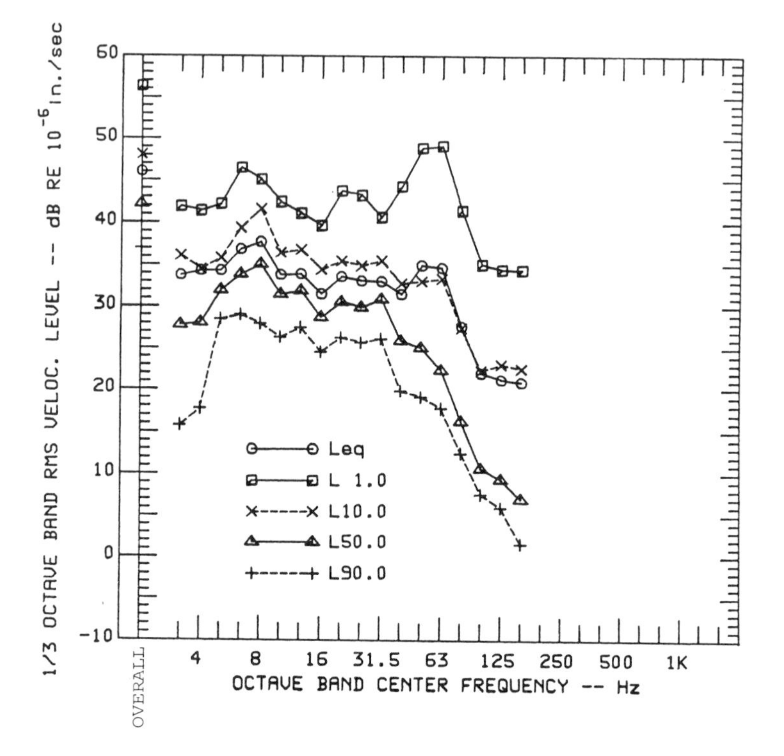

Figure 2. Statistical Analysis of 1/3 Octave Band Ground Vibration

In the case of existing buildings, the resonance frequencies of floors on which sensitive equipment will be placed should be determined. Basic techniques include performing a simple drop test with perhaps a 100 pound weight (a bag of cement or spare tire) or simply raising one's body up on the balls of the feet and dropping back on the heels, while simultaneously recording the velocity response of the floor. More sophisticated techniques may include measuring the input mobility of the floor at various points using a load cell and accelerometer or impedance head and an FFT analyzer (Reference 5). This latter approach, though technically difficult can provide data on the effective floor stiffness, mass, and damping factor at particular modal frequencies. Modal analysis can be employed to identify optimum locations for stiffeners and damping treatments.

In today's competitive semi-conductor industry, a thorough knowledge of the vibration environment is essential in maintaining the "yield" of most photo-lithography processes at an acceptable level. The type of measurements described above provide this necessary information. The data is more extensive than a simple seismographic recording but often is far more useful in determining appropriate vibration control techniques and evaluating potential sites with respect to criteria and equipment specifications.

Building Design

Foundation and building design play an essential role in minimizing vibration. Frequently, production and inspection machines are installed in areas that are inappropriate in terms of the vibration environment because other factors were considered more important. Examples include weak floor constructions, and close proximity to building mechanical equipment, loading docks, and freight handling equipment. The task of the building designer is to avoid these obvious pitfalls and, if possible, incorporate additional design features which will further reduce vibration levels. Some of these features include the use of deep friction pile foundations, high section modulus floors, floor vibration isolation, and vibration control provisions for building mechanical equipment, including laminar flow hoods used within the clean room area as well as other more conventional HVAC systems.

Effective foundation designs may include using deep friction piles for controlling foundation vertical response. Pile stiffness in the vertical direction varies with pile diameter and the square of pile depth (Reference 2). Increasing the pile depth is therefore effective in increasing pile vertical stiffness, thus lowering the response to incident ground vibration and vibration generated by equipment supported by the foundation. If bedrock is within reach, extending the pilings into the rock will provide a very high pile vertical stiffness and low response to vertical vibration. Finally, pilings designed for vibration control will be capable of supporting substantially more weight than that represented by the building's static load.

In the microelectronics industry, floors are being designed with high section modulus to achieve a relatively high floor natural frequency on the order of 30 Hz. Note, however, that for floors with a natural frequency of about 30 Hz, vibration caused by rotating equipment operating at 1800 rpm may create problems due to magnification near resonance. An effective approach to avoid this problem is to mount rotating equipment on floors which are structurally independent of the floors on which vibration sensitive machines are mounted.

In order to control floor vibration at the natural frequencies of the floor, sufficient damping must exist to absorb vibration energy. Part of this damping is provided by the concrete, part by radiation of vibration to other parts of the building, and part by partitions and equipment placed on the floor. This natural damping is usually sufficient.

For weak floors, dynamic absorbers have been considered for reducing floor vibration at resonance (Reference 1). A dynamic absorber is essentially a tuned spring-mass-dashpot system attached to the underside of the floor to absorb vibration energy at a modal frequency of the floor. Use of dynamic absorbers can be cumbersome due to their relatively high mass and difficulty in suspending them from ceilings in a safe manner. (The absorber's mass must be an appreciable fraction of the floor's modal mass to be effective.)

The most practical means for reducing the resonant response of weak floors is simply to add additional stiffeners or supports below the floor, thus raising the first bending mode resonance frequency and stiffness of the floor. This reduces the response of the floor to footfall induced vibration in the stiffness controlled region below the resulting first mode resonance frequency. This approach is usually more attractive than using damping systems which are effective only in the neighborhood of a resonance.

The floor on which vibration sensitive processes are to be located can be mechanically isolated from adjoining floor sections and independently supported on separate columns. Usually, a 1 inch thick gap with rubber seal is adequate. Figure 3 illustrates impact vibration reductions measured for a high stiffness waffle slab isolated floor construction used in the semi-conductor industry. The floor was independently supported by columns and isolated from the remainder of the floor. The vibration reductions were determined by impacting the unisolated floor at several distant locations and measuring the transfer functions between vertical floor acceleration at either side of the perimeter isolation joint at three locations. These data indicate a successful application of building vibration isolation techniques in the state-of-the-art semi-conductor industry.

For slab-on-grade floors, perimeter isolation has little effect at frequencies below about 60 to 100 Hz, because the underlying soil controls the response of the floor. Concrete slab-on-grade floors

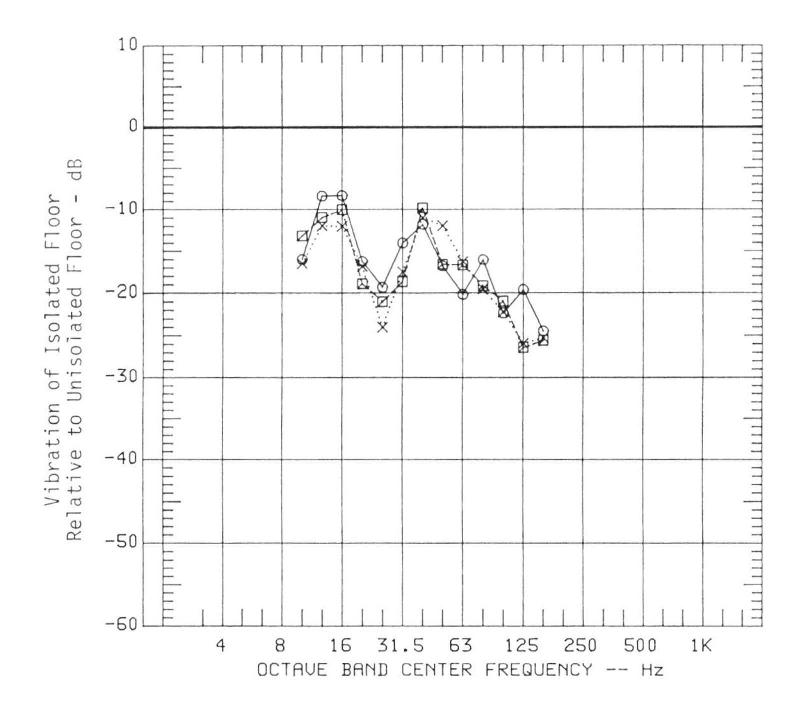

Figure 3. Vertical Vibration Reduction Effectiveness of Waffle Slab Floor Isolation
- Measurements at Three Positions -

are usually more desirable, however, than upper storey floors, since the response of slab-on-grade floors to vibration is much lower.

Vibration isolation of mechanical equipment, including HVAC equipment and ducts, elevator equipment, air compressors, and other equipment is very important for controlling vibration transmitted to sensitive areas. In this regard, the usual practices used for any large building are appropriate. In some extreme cases, air bags or pneumatic isolators and large inertia bases have been used to reduce low frequency vibration transmitted to sensitive areas. This latter approach is more sophisticated and requires auxiliary air pressurization systems. In addition to vibration isolators, care must be taken to provide flexible electrical conduit, piping, and ductwork to reduce vibration transmitted to the building structure.

After a building has been designed to minimize vibration transmitted to sensitive areas, additional vibration reduction can be obtained by locating sensitive equipment on isolation tables. Often, the equipment is supplied with its own isolation system. However, with proper building design the demands on equipment isolation systems will be substantially reduced,and the possibility exists of using relatively inexpensive elastomer or fiber glass pad isolators in lieu of expensive pneumatic isolators which must be carefully adjusted and maintained to avoid mechanical shorting.

Using elastomer or fiberglass isolators for isolating vibration sensitive equipment results in a higher resonance frequency for the isolation system than would be obtained with a pneumatic isolation system. The result is, perhaps, a lower settling time for the isolation system in response to internal machine stepping motions used during wafer or reticle inspection processes, thus increasing throughput and reducing inspection costs. These latter potential advantages can be realized only if building vibration is reduced to low levels and first mode resonance frequencies of the floor are above 25 to 30 Hz.

Conclusion

Although the demands on the semi-conductor industry and other industries applying high-resolution imaging technology are increasing with each year, methods are available for reducing or eliminating the impact of vibration on these processes. For scientific concerns and industry engaged in exotic biological processes, crystal pulling, or sophisticated large scale optical experiments, the same methods also apply, and, in fact, are all the more important. The building designer or manager must provide a low, practical level of environmental vibration for sensitive manufacturing and laboratory processes. Conversely, the equipment manufacturer must be aware of the limitations of particular buildings and design equipment capable of tolerating adverse vibration environments. If vibration can be minimized, the demands on the equipment manufacturer are lessened, with a possible improvement in performance and, ultimately, better products.

Appendix

References

1. Allen, D. L., Swallow, J. C., "Annoying Floor Vibrations - Diagnosis and Therapy," Sound and Vibration March 1975 pp 12-17.

2. Barkan, D. D., "Dynamic of Bases and Foundations," McGraw-Hill Book Company, N.Y., 1962, pp 46.

3. Saurenman, H. J., Nelson, J. T., "Procedures for Prediction of Ground-Borne Vibration and Noise from Rail Transit Trains," (Technical Memorandum) Wilson, Ihrig & Associates, Inc., Prepared for U. S. DOT/TSC under Contract DOT-TSC-1796, October 31, 1984.

4. Ungar, E. E., White, R. W., "Footfall-Induced Vibrations of Floors Supporting Sensitive Equipment," Sound and Vibration, October 1979.

5. White, M. F., Liasjo, K. H., "Measurement of Mobility and Damping of Floors," Journal of Sound and Vibration 81 (4) 1982, pp 535-547.

SUBJECT INDEX

Page number refers to first page of paper.

AUTHOR INDEX

Page number refers to first page of paper.